CIM Implementation Guide

First and Second Edition Title:
A Program Guide for
CIM Implementation

Dr. Leonard Bertain
Editor

D1249443

Richard Perich
Publications Administrator

Published by the

Computer and Automated Systems Association
of the Society of Manufacturing Engineers
Publications Development Department
Reference Publications Division
One SME Drive
P.O. Box 930
Dearborn, Michigan 48121

CIM
Implementation Guide

Copyright © 1991
Society of Manufacturing Engineers
Dearborn, Michigan 48121

Third Edition

First Printing

Library of Congress Catalog Card Number: 91-60114

International Standard Book Number: 0-87263-400-0

Manufactured in the United States of America

CASA/SME

The Computer and Automated Systems Association of the Society of Manufacturing Engineers (CASA/SME) was founded in 1975 to provide comprehensive coverage of computer automation and integration for the advancement of manufacturing.

CASA/SME is applications-oriented and addresses the integration of all phases of marketing, research, design, operation, installation, customer support and communication as they affect manufacturing.

CASA/SME activities are designed to: (1) provide professionals with a single vehicle to bring together the many aspects of manufacturing, utilizing computer systems and automation as tools, (2) provide liaison among industry, government, and education to identify areas where computer and automation technology development can be beneficially applied, (3) encourage the development of totally integrated manufacturing.

Introduction

Successful implementation of CIM technology requires extensive planning as the first step in the process. The examples cited in this book achieved success because a great deal of thought went into planning how this technology would be applied and why. From these installations, we observe that the planning process can only work if it involves all personnel of the targeted business unit.

In this compendium of articles, we document the process leading to a successful CIM system. We are pleased to have the participation of Dr. Richard Schonberger, whose book World-Class Manufacturing has established a standard of excellence required of any manufacturer. In his article, he asks us to think about the meaning of the letters "C" and "I" as they relate to CIM. He raises some interesting points for us to consider. One that is especially dear to my heart is: "keep it simple." This concept leads to Warren Shrensker's "A Brief History of CIM," from which we establish a reference point for later discussion.

I have tried to organize the remaining materials into six parts:

Part One: Strategic Thinking

Part Two: Tools and Technology of CIM

Part Three: CIM Implementation Results and CIM Survey

Part Four: Japanese CIM Installation

Part Five: The Product Data Exchange Standard

Part Six: CIM Survey

In each of these, we find some good content for review. My own article entitled "Productivity and CIM," discusses the results of productivity improvement by getting the personnel of the organization to understand the basic principals of good manufacturing. CIM systems follow naturally after the initial analysis of the business is completed, and people have been able to take action. Mike Roberts and Lew Soloway analyze the steps required to sell CIM implementation on a unique cost justification process. It is mandatory reading for anyone interested in trying to justify the cost of such a system. Tom Billesbach looks at the planning steps necessary for organizing installation.

In the section of the book entitled, "Tools and Technology of CIM," the normal tools of the trade are discussed quite ably by the contributors. The thoughts of CAD/CAM, group technology, IDEF charts, communication standards, and tool room optimization are all excellent presentations. The article by Jim Johnson, discusses a new way of analyzing the work process using his idea of EventMap (TM) diagramming. As a practitioner of this method of diagramming, I know it works.

Finally, we have several success stories from past CASA/SME Lead Award Winners. We thank Dr. Yasuo Mizokami of the OKI Electric Industry Co. of Tokyo, Japan for sharing his thoughts on the very successful CIM implementation of a semiconductor plant. We also appreciate the contribution from Fred Michel and Mark Pardue on surveying world-class CIM planning. Thanks also go to Robert Carringer for his article on the Product Data Exchange Standard, and to Gary K. Conkol for contributing information on "The Role of CAD/CAM in CIM."

Numerous other contributors also helped put this document together. The Editor wishes to thank Mr. Robert King, and Mr. Richard Perich of the Society of Manufacturing Engineers' Publications Development Department for their support during the preparation and delivery of the final materials to the printer.

Dr. Leonard Bertain, Editor
American Productivity Group
Oakland, California

Table of Contents

PART SIX -- CIM Survey

General Comments

The "C" and the "I" in CIM

By Richard J. Schonberger
Schonberger & Associates

INTRODUCTION

I had just finished speaking at a seminar in London, England, and consultant Paul Collins was next. Paul told the audience, "There is nothing wrong with CIM--except perhaps the 'C'." I grinned and nodded.

What Mr. Collins meant, and I agree, is that the first priority in CIM implementation is integration of processes--wholesale movement of people and equipment out of functional departments and shops and into many focused groups. Computer-integration comes later, where necessary.

FUNCTION-BUSTING

This integration, this creation of product or customer dedicated cells and flow lines as well as the geographical relocation of resources (people and all their equipment and accessories) applies to both factory and office. At the beginning of the 1980s, factory automation was number one on everyone's industrial revitalization list. By the end of the decade, it was clear that moving machines into cells had been the far more popular activity, and justifiably so.

The other part of revitalization involving fixing the bloated, error prone, time wasting support functions, did not take the same path. There has been some excellent work done recently in creating cross-functional design teams (1), as well as office cells to process orders. But office automation has been the more common pursuit. Under the office automation formula, everybody stays in their functional departments--order entry, accounting, scheduling, materials, engineering, personnel--and does their planning and controlling from one PC to another. Given the advanced state and relatively low cost of information and messaging systems, is there anything wrong with that?

Yes. It is a sure formula for errors, delays, high costs, and overall poor performance. Communications and information processing is not the problem. The problem is the gang mentality that goes with the territory of the functional organization. Computer linkages across space help bridge distance; still, the parties on each end of the line owe first allegiance to "their own kind," not the customer.

SEARCHING FOR THE CUSTOMER

The late Kaoru Ishikawa is the originator of the elegant idea that the next process is your customer. The idea is to link yourself through a chain of customers to the final user. But if the next process--factory or office--is in another department and distant in time, geography, and organizational goals, you are not going to be focused on the customer.

Some years ago, sort of with tongue in cheek, I suffered my seminar audiences to look at an overhead transparency labeled "Schonberger's Law of Effectiveness." I had since forgotten about it, but recently learned that a few companies were promoting it. It goes like this: the effectiveness of an activity is inversely proportional to its distance away from the primary mission. In this case, distance refers to the geographical organizational process.

The real success stories are in the firms that not only move their technical support people into focused teams but also move them out of central offices and close to their focused counterparts on the plant floor. For example, Physio Control, a producer of defibrillators, broke up all of its plant processes and reorganized into eleven product focused cells (2). Furthermore, it assigned a "response team" of five people to each cell. Most had been relocated to or near the plant floor. Each response team includes a buyer, a process engineer, a sustaining engineer, a quality engineer, and a test engineer. They do not have quite enough to go around, so a buyer may serve on three different response teams. The industrial and business press recently has devoted several cover stories to the same idea, citing many examples of extended design-build teams.

AWARDS

In 1981, the Society of Manufacturing Engineers (SME) along with its subsidiary, the Computer and Automated Systems Association of SME (CASA/SME), initiated the LEAD (Leadership and Excellence in the Application and Development of Computer Integrated Manufacturing) Award: The implied premise was that computer-integration was inherently good, a widely held opinion at the time. Today, I believe that opinion is changing to functional integration is a necessity; computers and automation are options.

What about those options? What is the role of computer automation, once we get functional integration correctly placed as first priority? Computer automation is mostly for making value adding processing improvements; examples are computer-directed metalcutting and assembly. Its secondary, sometimes temporary role is integration of far-flung resources. For example, passing design information back and forth and synchronizing schedules from plant to plant. To the extent that geographical merger or simple, visual kanban are feasible, these computer linkages are best viewed as temporary; progress is removing them.

Someone wrote: "simplify, automate, integrate." Let's fix that statement: first integrate the processes, which is the greatest simplifier ever devised. What comes next falls into place naturally.

Does the LEAD Award need to be refined, restated, or junked? I do not know. What I do know is that it is folly to simply make awards to technological wonder children. What needs to be praised and encouraged are those who can become the lowest cost, highest profit, lean-and-mean competitors in their industry. In my mind, that points to the plant or company that has thoroughly taken itself well down the path of functional integration and is using computer automation very selectively to drive for perfection in the eyes of the customer.

(1) Peter Nulty, "The Soul of an Old Machine," Fortune, May 21, 1990, pp. 67-72.

(2) Richard J. Schonberger, Building a Chain of Customers: Linking Business Functions to Create the World Class Company, New York: The Free Press, 1990, pp. 53-54.

A Brief History of CIM

By Warren L. Shrensker, CMfgE

General Electric Company

INTRODUCTION

The need to improve productivity, as well as product quality, and reliability while cutting costs, cannot be met by just better equipment, skilled people and/or islands of automation. There must be better managerial tools and philosophies, and integration of various business disciplines to define, produce, and market products that take advantage of existing resources, equipment and people. The one tool that can meet these challenges in a cost-effective manner is computing systems technology.

COMPUTER TECHNOLOGY EVOLVES

Computers are not a recent innovation, although many people are now experiencing them through personal computers. The first computer was created around 1946. A commercially available computer did not enter the marketplace until 1951. The first digital computers were used by universities and the U.S. government. It was not until November 1953, that the first digital computer (UNIVAC I) went into commercial use. It was leased by the General Electric Company from the Remington Rand Corporation for use at GE's Appliance Park in Louisville, Kentucky for financial applications. In 1954, numerical control (NC) was introduced. Then in 1955, development of the first automatically programmed tool (APT) processor marked the premier of computer-aided manufacturing (CAM). Today, CAM encompasses many more manufacturing disciplines. Computer-aided design (CAD) appeared publicly early in the 1960s in high-technology design businesses (automobile and aerospace) with design augmented by computers.

In the late 1960s and early 1970s, with the advent of microelectronics, came the minicomputer, and more importantly, the microcomputer--the "computer on a chip." The performance of these systems was excellent. Costs continually dropped and systems found their way into more manufacturing applications and into computer-aided design products, such as interactive graphics. In 1974, Dr. Joseph Harrington coined the term "Computer-Integrated Manufacturing" (CIM) in a book he published by the same name. However, it was not until 1981 that the term became widely used.

A major boost to the CAM technologies actually came from the Department of Defense when in 1975, they started the AFCAM (Air Force Computer-Aided

Manufacturing) program which was intended to erect a scientific approach to better manufacturing technology. Out of this in 1976, the ICAM (Integrated Computer-Aided Manufacturing) program was born under the direction of the U.S. Air Force's Materials Laboratory and under the guidance of the National Academy of Engineering's Committee on Computer-Aided Manufacturing (COCAM). During the mid-1980s the ICAM program was changed to the Air Force's CIM program and then later changed to the parent MANTECH (Manufacturing Technology).

In the 1980s, CIM capabilities grew at a rapid pace. Chip technology continued to improve, from 8-bit to 16-bit to 32-bit with cost/performance ratios improving annually. This drove computer-aided design and manufacturing systems to new limits. Computer-aided design had grown in the early 1980s into a new discipline called computer-aided engineering (CAE), which utilized graphics with powerful analysis programs, including solid modeling. This was joined by a wide range of other computer-aided systems for manufacturing process planning, shop scheduling, inventory control and decision support. More powerful computers were constantly being developed to support these applications, like the super minicomputer, supercomputer, personal computers, engineering work stations, programmable logic controllers, and even the use of general purpose mainframe computers with transaction processing and vector processors. Computer networking allowed computers to pass data to one another so systems could be integrated.

THE FUTURE OF CIM

Industrial technologists of the world have forecasted that the overall future trend in engineering and manufacturing beyond the year 2000 is toward the development and implementation of computer-integrated manufacturing. Significant economic and social incentives are at work to provide the motivation for this to happen. The strategy being followed is to develop and implement a sequence of viable economic steps in shorter-range programs to bring about the eventual realization of the overall objective. These involve development and implementation of new optimization technology with integrated engineering manufacturing databases, group technology, cellular systems (including automation and robotics), and full manufacturing management systems and their applicable complex software systems. They would also include the latest development in graphics, engineering analysis, business systems, and office automation, etc. As we look into the 1990s and beyond, artificial intelligence (mostly expert systems) is being wrapped into these types of systems.

The Computer and Automated Systems Association of the Society of Manufacturing Engineers (CASA/SME) has spearheaded the use of the term computer-integrated manufacturing (CIM) that provides computer assistance to all business functions from marketing to product deployment. The term "manufacturing" in CIM refers to the manufacturing enterprise with all of its disciplines, not just the disciplines of engineering and manufacturing. Marketing, sales, distribution, finance, human resources and general office administration are also included. Computer-integrated manufacturing (CIM) embraces what historically has been classified as "business systems" applications. This includes order-ship-bill, bill of material development, manufacturing resource planning, design automation, drafting, design and simulation; manufacturing planning, including process planning routing, tool design and numerical control parts programming; also shop floor applications, such as numerical control, assembly automation, testing, statistical process or quality control, and process automation. It also covers any other administrative and financial systems required for the manufacturing enterprise to function efficiently. In other words, through communication, it is getting the right information to the right person at the right time to make the right decision.

A fully integrated CIM system involves the design, development, or application of each of the systems in such a manner that the output of one system serves as the input to another. For example, at the business planning and support level, a customer order servicing system receives, from the sales force, descriptions of products to be purchased by prospective customers. The product description serves as an input to the engineering design function. If the product contains previously designed components, a computer-aided design system would output the engineering drawing information to the bill of materials processor and process planning system. If the product description contains new components, the description would serve as input to a computer-aided design system where the graphics could be used as a design aid to provide engineering and manufacturing information. Complete implementation of CIM results in the automation of the information flow in a business organization from order entry through every step in the process to shipment of the finished product and invoicing and payment receipt. People often ask, "Where can I buy a CIM system that does this?" You cannot buy a total CIM system. CIM is a philosophy. This philosophy allows a corporation to develop their systems into CIM.

As corporations started to implement CIM systems, it was noticed that problems with implementation were not so technically hindered as they were people hindered. This realization opened a new area to explore called management science. CASA/SME started to explore a new management concept that would allow CIM to succeed. It was called "Fifth Generation Management." It was spearheaded by Dr. Charles Savage, a member of the CASA/SME board of advisors. Through a series of Round Table discussions (two CASA/SME Blue Books were published in 1987), presentations with audience participation and further research, Dr. Savage developed his theory of Fifth Generation Management, the management organizational structure and methodology that will be required to compete in the Nineties and to manage the enterprise that has or is implementing CIM. We now know that to successfully have a computer-integrated enterprise, there are two aspects of computer-integrated manufacturing; one involving technology, and the other involving management style, structure and people.

Part 1

Strategic Thinking

Productivity and CIM

By Leonard Bertain, PhD
American Productivity Group

INTRODUCTION

Several years ago, I was asked to design a system for a large company. It was a very sophisticated vision system for inspecting a high-speed line of nonwoven material goods. It required the design efforts of the manufacturing engineering department, electrical engineering department, the industrial engineering department and miscellaneous interested parties. In each meeting that I recall, there were no less than 15 people in attendance from this company.

In each case, however, one party was conspicuously absent.

This did not occur to me until many years later. After the system was installed, it had one remaining glitch; it just wasn't working to the specifications of the design.

Tensions were building in the plant. The plant manager was beginning to get worried. When he worried, everyone worried. As I stood by the system one day while it was still inoperable, I could see that there were over 13 engineers from the company who were worried. And so was I. If the system didn't work, we would not get our final payment.

As I stood back from the machine about 50 feet from the line and leaned against a wall, one of the ladies who had been working the line since its installation approached me. She asked me what the problem was and I told her. She thought about two seconds and pointed out something that no one--including me--had thought of. As I approached the crew with this new information, I wondered: Should I give credit where credit was due? I approached the group with the suggestion. No one was willing to accept the fact that this good idea had been proposed by one of the line workers.

I never really gave the issue much consideration until recently. I have been doing productivity improvement implementations at several companies in the San Francisco Bay Area. In each case, the projects started with a solicitation of ideas from employees. Implementation of those ideas worthy of action served as the basis of a system that leads directly to a significant optimization of the work in the factory.

The missing "participant" I referred to at the beginning, was none other than the line worker who knows the process. These production workers are being recognized as the key ingredient to the design and implementation of any successful CIM system. Most consultants in the industry would agree with this observation. The message is clear to anyone beginning a CIM Implementation: "Get everyone involved. Don't leave anyone out of the planning process."

Successful CIM implementation has been the result of one of the projects and will be the ultimate conclusion of the other six projects discussed here.

I find good, sometimes even brilliant, ideas originating from production workers and clerical staff at each company that has been part of these efforts.

The subject of productivity and CIM begins with the issue of recognizing the importance of production workers in the process of developing a CIM system. As Dr. Richard Schonberger observes in his article in this publication, "simplify, automate, integrate." I would suggest that the process is: "motivate, innovate, automate, and then integrate." Aristotelian logic, therefore equates this to mean that to simplify is to motivate and innovate. The rest of the article will deal with the issue of simplification by a process of motivation and innovation. The system which has evolved from these efforts is called VAMMS (Value Added Management, Manufacturing and Service).

THE VAMMS SYSTEM

Tom Peters has been very successful at chronicaling the stories of companies that have been successful at implementing employee involvement programs. Harley-Davidson, North American Tool and Die Company, and others are well documented success stories. Executives have shown their support publicly and internally and have made the commitment to make these programs successful.

We were not so lucky. As we started the education process at the first installations, ideas began flowing. Some were implemented and everything looked rosy. However, one day the ideas stopped. People didn't care and the spirit of enthusiasm for class stopped almost immediately. Why?

Like all good analysts, we probed the organization and found the first flaw in the original concept: top management was giving the concept of employee involvement lip service but not commitment.

In future implementations, we got that commitment and the organization began to respond to the training process. The training began with the basic idea illustrated in Figure 1. This ensures that no one in the training process is blamed for suggesting that someone else's idea or process area can be improved. In fact, once the blame issue is addressed, the program can proceed quite nicely. The blame issue is probably more directly involved in the reporting of information. If people get yelled at for giving managers "the truth" instead of the data that they need to report, the system will not work. We have mastered, in this country, the art of shooting the messenger of bad news. Most American managers spend a lot of time reviewing reports which are tainted with inaccurate information. Decisions are made on this information and companies spend long hours adjusting business decisions made with this inaccurate information. In essence, all aspects of the data collection

Figure 1. "NO BLAME" ensures a smooth progression of the training process.

process in business must be based on the assumption of "NO BLAME." As Jack Webb said, "Just give me the facts, ma'am."

The concept of "NO BLAME" originally was prompted by an article in Manufacturing Engineering magazine of February, 1987, an excerpt of which appears in the following paragraphs:

Konosuke Matsushita, founder and executive adviser for Matsushita Electric Industrial Co., Osaka, Japan, is certain that his country will win the manufacturing war. Not only that, but he thinks he knows just why that will be. "We will win and you will lose," he says. "You cannot do anything about it because your failure is an internal disease. Your companies are based on Taylor's principles. Worse, your heads are Taylorized too. You firmly believe that sound management means executives on one side and workers on the other. On one side men who think and on the other side, men who can only work. For you, management is the art of smoothly transferring the executives' ideas to the workers' hands.

"We have passed the Taylor stage. We are aware that business has become terribly complex. Survival is very uncertain in an environment increasingly filled with risk, the unexpected, and competition. Therefore, a company must have constant commitment of the minds of all of its employees to survive. For us, management is the entire work force's intellectual commitment at the service of the company...without self-imposed functional or class barriers.

"We have measured--better than you--the technological and economic challenges. We know that the intelligence of a few technocrats--even very bright ones--has become totally inadequate to face these challenges. Only the intellects of all employees can permit a company to live with the ups and downs and the requirements of its new environment. Yes, we will win and you will lose. For you are not able to rid your minds of the obsolete Taylorisms that we never had."

Editor's note: Frederick Winslow Taylor (1856-1915) propagated the principles of "scientific management" in industry and business.

Matsushita makes it clear that American companies will never be able to compete against the Japanese because the American managers' heads are "Taylorized." They are not able to allow subordinates the freedom of input into the organization. Only the managers' ideas are acceptable.

Workers are willing to contribute to the improvement of every factory in this country, but they need to believe their ideas will be acted upon and they need to be assured that there will be no repercussions for offering ideas that counter those of their line supervisors. Believe me, the American worker is smart enough to test both of these issues at the very beginning of any project.

We have seen it in every project that we have begun. We usually see some form of testing by the end of the second day of training. Probably one of the most dramatic examples of this was recently provided by a member of the union in one of our recent projects. We begin all projects discussing waste in the business. We spend about a week going over the different areas of waste. In a typical company we find about 50 to 100 identified wastes per class per week. In this process, different workers get very motivated by the class activity and immediately want to take action. In this particular company, one of the production workers asked me if I really believed that management was going to change. I told him, "absolutely." He went home and over the weekend solved one of the waste issues on his own time. The production worker generated a five-page written proposal because he was excited about the opportunity of making a difference. The idea was going to reduce a paint transfer changeover from five minutes to 90 seconds for a total cost to the company of about $300. In fact, he was so excited that he turned in his idea to his supervisor who promptly threw the paper into the waste basket, exclaiming "It'll never work!!"

Needless to say, the damage was done. It took another two weeks to get everyone excited again. This time, everyone was on guard. Executives were slow in getting the commitment from others in the organization and the productivity improvement cycle was stifled. Management must give a commitment to the process of change. They must also convey their total commitment to all of their line managers and make it clear their support is needed to make the change successful. (The Bertain Corollary to this is: you're in or you're out.)

During the VAMMS process, employees are trained to identify wastes. They are instructed to look at 12 categories of waste, including those of the Toyota Production System and several others. Then they are trained to analyze the wastes, look at the various causes of waste and finally propose solutions. In its simplest form, the VAMMS process is a problem-solving model. It uses all the concepts of SMED, JIT, Kanban, Push vs. Pull system, etc. that are helpful tools in analyzing problems of the workplace, as shown in Figure 2.

Of all the tools we have tested in the workplace, the most interesting is described in the article by James Johnson in this publication. He has developed a very useful tool to describe the processes that take place in any business. He calls the product EventMap (TM) Diagrams. You can read the article to get a better understanding of the concept. I mention it here, because it is very useful in designing a CIM system.

We have used it extensively in all of our projects. It is easy to understand, and is a very powerful tool in laying out designs, describing the

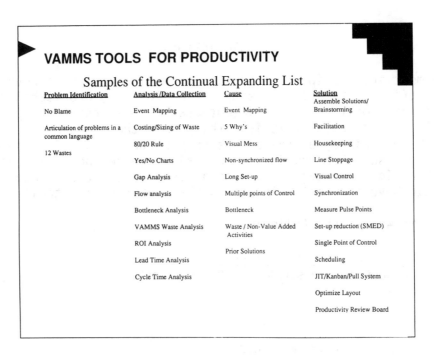

Figure 2. The VAMMS process.

paper flow process of a business, or even the process flow of a doctor's office. Fred Michel's article points out that one of the biggest problems identified by companies in implementing CIM systems is communication. Most organizational effectiveness consultants also cite this as the most significant problem in industrial organizations. The EventMap (TM) process has proven most useful at improving those skills. Instead of people yelling in frustration at each other when they are trying to convert colleagues to an abstract principle or vision of a process, draw an EventMap (TM) of the process. It may even save lives.

We use it as an ongoing tool to describe the cash flow time line. For us, this time line consists of things that go into the process of converting an order into cash. Try it in your organization. If you have any problems, give me a call at: (415) 763-8871 and I'll put you in touch with James Johnson.

Once the people in the business learn the skills of problem solving proposed by the VAMMS system, the key measure is the degree of success of the resulting continuous improvement and employee involvement system. We use the ideas generated as one measure of this success. If you are getting a steady stream of ideas, everything is probably all right. When the ideas stop, the system has stopped performing. But the system is not as simple as that. We usually require the implementation of a productivity improvement program to monitor the progress of the company in productivity measures. The system of monitoring this process requires the total commitment of company executives. This ensures a systematic process of measuring productivity. Goals need to be established and the "Productivity Czar" (or Productivity Champion) needs to be able to make commitments to program improvements, including a discretionary budget.

Once this system is in place, it will quickly lead to an improved manufacturing process. Automation and integration are natural steps after the process of simplification: simplicity equates to motivation and innovation.

OBSERVATIONS

One of the most interesting consequences of motivation and innovation is increased profitability. In the simplest definition, profitability is defined as the ratio of profit to the efficient use of those assets to produce those profits. In many circles, this is nothing more than return on assets. By analyzing the efficiency of the assets to produce the profits, setup time becomes a big issue when you look at a small machine shop. If the cost of the machines is a big percentage of the total asset base, then setup time reduction can have a dramatic impact on the measured profitability.

In these terms we see the following:

$$\text{Profitability} = \frac{\text{Profit}}{\text{Efficient use of assets to generate the profits indicated}}$$

$$\text{Profitability} = \frac{\text{Revenue} - \text{Costs}}{\dfrac{\text{Equipment} + \text{Inventory} + \text{A/R} + \text{Cash}}{\text{Efficiency}}}$$

In this equation the equipment is really the building and equipment. Clearly, the efficiency of setup can have a dramatic effect on profitability. Inventory, of course, consists of all these components: Work-in-process (WIP) + Raw materials + Finished goods. In summary, the profitability equation serves as a good illustrator of the importance of all the ideas of productivity improvement. Wastes of excessive WIP lead to reduced profitability.

Over the last 50 years, we have focused all our energies on cost containment, which is just one component of the profitability equation. In fact, very few accountants know how to deal with the issue of profitability or return on assets presented in these terms. The article by Mike Roberts and Lew Soloway in this publication opens up the discussion of these issues from a different perspective, namely the wise investment of capital. They discuss the CIM investment from a new set of criteria. By wrestling with these new issues of investment strategy in the course of training, the financial organization of the company is able to lead a number of discussions regarding the best use of capital to serve the customers of the company.

Figures 3 and 4 show two methods of analysis that have proven useful. By getting the employees to understand the Gap Analysis, they are able to relate company capabilities to customer requirements. Where these two overlap, there is value. The easiest way to do this type of analysis is to let the production workers and staff talk to a real customer.

In Figure 4, cost benefit analysis and mission relevance are considered in the training with the idea of creating an understanding of those issues that will make the company successful.

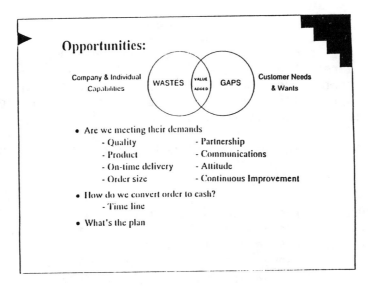

Figure 3. Gap analysis.

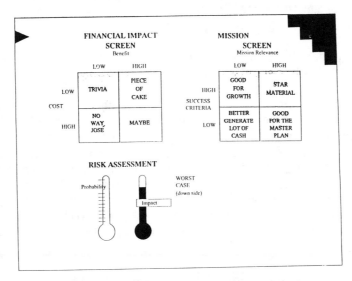

Figure 4. Cost benefit and mission relevance analysis.

Once these issues have been thoroughly discussed, the workers will have a better understanding of why they are working and what the goals of the company are. The employees then drive to take complexity out of the system. They are firmly motivated to improve profitability because they see something in it for themselves. Management has got to take care of the workers and not just stuff their own paychecks. Workers will continue to be innovative in improving the system if a strong measuring system is established that eliminates blame for failure, but encourages the risk of creative input.

CASE STUDIES

Case 1

The first case was a program at a small Bay Area company which remanufactures starters, alternators, and generators. The main waste in the business was looking for parts. About 75% of the assemblers' time was spent looking for parts. During the training, about 230 ideas were generated in two weeks from 30 employees.

The main idea was to create a series of dedicated cells for each of several different product lines. The net throughput of the individual operator improved six-fold as a result of creating a cell. The process of plant improvement has been continuous. Before all the other cells can be implemented, large quantities of junk "cores" had to be inspected prior to being thrown away. This was a very tedious process.

The benefits of the training and subsequent cell layout might be summarized as follows:

1. Lead time improved six-fold.
2. Inventory has been reduced 50% in the first year alone and is expected to be reduced another 70% before the second year is over.
3. Worker skill level has been increasing.
4. Workers who did not speak English very well have had to improve their language skills (with classroom training) and managers who did not speak Spanish are learning.
5. Workers have a greater say in their work environment.
6. The company is making greater profits by a significant amount over its previous poor performance.

More important than the success of this project is the subtle management and organizational problems that have surfaced:

1. The company was started by the owner when he was very young. In the meantime, he has had to give up control of some information that is needed to run each cell. Each cell is a separate business unit. The cell supervisor needs to have better knowledge of the raw materials or "cores" that make up his business. As the first cell was implemented, the owner felt the loss of control. As each new cell was implemented, more and more of his control was disappearing. Upon reflection, every company that we have worked with has had this issue arise. CIM systems require a complete distribution of work and information.

   However, the management theory of change says that as control is moved down to the cell, managers and supervisors need to learn new jobs or different control measures. Our simple approach is that work consists of three things: plan, control and do. Most of the workers that we have dealt with merely "do". The managers get to "plan and control." After the VAMMS process, the workers in the cell get to do the whole process: plan, control and do. So, the management dilemma becomes: what is my new planning and control activity?

2. The process of implementation went slower than I would have ever expected. Change is a slow process but the change in this operation was really slow. It involved not only a change in management philosophy, but a real culture change in the company.

A final observation: As the company is going through a restructuring of its information and management, everyone was affected. This simple manufacturing improvement project has led to phenomenal productivity improvement. It is well on its way to a complete integration of factory activities directly with the customers of the company. The CIM activity in this shop has been a pay-as-you-go project. There have been no major expenditures of capital without clear benefits to all members of the organization.

Case 2

The second case is a small machine shop which is a JIT supplier to a Bay Area semiconductor equipment manufacturer. The company was profitable but the owner was trying to improve the operation.

It was clear from initial discussions that setup time was a major problem. In the two weeks of brainstorming roughly, 130 ideas for waste elimination were input from 35 employees. The value of the wastes was identified to be $1.7 million. Setup accounted for 25% of the N/C operator's time. Prior to the training, the owner recognized that a "Ball-Lock" quick changeover system would help reduce overall setup time. The concepts of SMED (Single Minute Exchange of Dies) were made clear to everyone in the company, including the accountants and receptionist. The process of setup reduction went very quickly once the "Ball-Lock" system was in place. Interestingly, the owner made a decision to go with the new system, and it was a total commitment on his part to make it succeed. He knew it would work, he was just making sure that everyone understood. The results are telling. Setup is down dramatically. Profits are up. Approximately 30% more work is being generated with the same number of people as last year.

The customer is getting ready to start sending drawings and N/C machine programs over communications lines before the end of the year. The CIM program is defined and ready for implementation and was the product of everyone's contribution.

Benefits can be summarized as follows:

1. Setup time fell to less than 10%. The company is actually able to do more setups on smaller runs.
2. Inventory is down by about 70%. The inventory level is controlled by a kanban system whose reorder level is controlled by the floor personnel.
3. Profits are up.
4. The processing time for each job is down. Because there are more setups, the owner decided to add another programmer to the operation. He created two cells and each programmer controls the operations in each cell. By giving the programmers more time to focus on each cell, they have been able to generate better quality processing. On top of this, the programmer was moved physically closer to the cell area.

Problems noted during the implementation:

1. The greatest concern of productivity improvement systems is having workers learn to cope with periods of inactivity. With improved performance, there will be free time. As long as the free time does not lead to layoffs, this is a time when new skills can be learned through cross-training or other skills training.
2. In this shop, as the productivity and performance of each cell improved, work was moving through the shop so fast, there was no noticeable back-up inventory laying around the floor. Most workers are trained to feel comfortable with lots of material stacked up. There is a feeling of discomfort if there isn't a lot of inventory around. An excessive amount of inventory means that there is work to be done. In the past, if there was no material sitting around the

shop, a layoff would soon follow. In order to make JIT systems work, a set of different feedback measures need to be provided to ensure that jobs are secure. In this application, the backlog of committed orders was made available to the workers to lower anxiety.

OTHER CASES

Similar projects were developed at a plant that makes 55 gallon drums, a company that supplies purge panels for the semiconductor industry, a company that manufactures valves and regulators for semiconductor purge panels, and a plant that makes diesel control systems for stationary diesel systems.

In each case, ideas were generated, analyses performed, causes determined and excellent solutions proposed and implemented.

CONCLUSION

If there is a single observation that can be made about improving the productivity of a company, it is that technology is not the answer. It may be part of the answer, but it is not the only answer. People play an important role in getting to the real issues that limit productivity in plants in America.

Some good advice to follow is offered by Tracy O'Rourke, chairman and CEO of Varian Associates, in Palo Alto, CA. In an article in Manufacturing Engineering several years ago, he was quoted as saying: "I have 10 simple rules for running my business: rule number one is find a simple project and implement it. Rule number two is find a simple project and implement it. Rule number three is like the first two: find a simple project and implement it. Rules four through 10 are just like the first three."

In beginning any project, establish the objectives and implement the project according to the 80/20 Rule. It will be more cost effective and will get you to your ultimate goal more quickly.

I have taken Mr. O'Rourke's rule as my final presentation in Figure 5. Follow it and you will succeed. It is the foundation of continuous improvement and CIM systems.

Find a simple project
and just do it

Find a simple project
and just do it

Find a simple project
and just do it

Find a simple project
and just do it

Find a simple project
and just do it

Find a simple project
and just do it

Find a simple project
and just do it

Find a simple project
and just do it

Find a simple project
and just do it

Figure 5.   The foundation for continuous improvement and CIM systems.

CIM, CALS and Knowledge Networking

by Dr. Charles M. Savage
Digital Equipment Corporation

INTRODUCTION

After 27 years, we are on the verge of realizing the full implications and benefits of Joseph Harrington's vision of computer-integrated manufacturing (CIM). We have been slow in understanding that CIM involves more than just technology. It also involves the way we work together as professionals, impacting both technology and organizations.

Why has it taken so long? Harrington's 1973 book, "Computer-Integrated Manufacturing," foretold of the removal of the slash between CAD and CAM (CAD/CAM). He understood that the power of the computer could bridge the traditional gap between engineering and manufacturing.

In 1984, Harrington published a second book, entitled "Understanding the Manufacturing Process." In this work, there is remarkably little mention of CIM. Instead, he argues that the manufacturing process is an "undividable tapestry." Perhaps Harrington suspected that the discussion of CIM would get bogged down on the shop floor, just as it has. He wanted to broaden the discussion to include the organizational aspects of integration. He must have envisioned the development of CALS (Computer-aided Acquisition and Logistics Support).

"BIG" CIM

CALS is really "big" CIM. It challenges us to think not only in terms of CAD and CAM, but also of logistics, technical documentation and well-architected data repositories. CALS is quickly spreading its influence beyond the defense community, where it began. Discrete manufacturing, repetitive manufacturing, and even process industries will benefit from "big" CIM.

Those developing CALS have wisely recognized that linking computer based technologies is only one-third of the challenge. Our real task is to move to "Concurrent Engineering" (CE). Two-thirds of our energy is needed to learn to work organizationally in new ways.

To learn new ways of working, we must unlearn many of the basic assumptions of the industrial era. We cannot expect to "empower" ourselves if we do

25

not understand how "encaged" we are in the attitudes, values and norms of
this industrial age which include:

* Division and subdivision of labor leading to unnecessary
  fragmentation
* Pay for narrowly defined tasks which does not respect human
  knowledge and capabilities
* Petty self-interest which leads to "turfitis" and stovepipe
  management

Soon, we will be able to put 50 to 100 MIP work stations on the desks of our
design engineers, manufacturing engineers, and logistic engineers and
network them with 100 to 300 MB fiber optics. Yet, without trust and
respect for people's capabilities and knowledge, the investment is hardly
worth the money.

KNOWLEDGE: THE NEW SOURCE OF WEALTH

As we enter the knowledge era, we find our source of wealth shifting from
labor and capital to the knowledge and capabilities of workers,
professionals, managers, and executives. This represents change with
profound human implications.

How are we to tap into this motherlode of wealth? How do we build on one
another's capabilities? How do we grow little ideas into well-targeted
quality products and services? How do we configure and reconfigure multiple
task-focusing teams? How do we develop "teamwork of teams" among and
between multiple concurrent engineering teams?

Most of the key ideas and notions needed to develop our businesses are still
in people's minds, not in applications programs and databases. Therefore,
our biggest challenge is to network this knowledge. "Knowledge networking"
assumes we will finally make it easier to work in a collaborative mode,
building upon one another's thinking. We might be helped by powerful
computers and high throughput networks, but the real breakthroughs will come
as we adjust our norms and values.

Breaking through these conventional standards will not be easy. They are
ingrained in us from an early age. Since early youth, many of us have
developed our ability at noticing one another's weaknesses. When we knew
the answer to the teacher's question and our classmates did not, we were
secretly delighted. Over the years, we have honed our skills at exploiting
and manipulating each other's weaknesses.

It is little wonder, then, that design engineers want to complete every
detail of the product design before handing it over to the manufacturing
engineer. There should be no "weaknesses" in the design against which the
manufacturing engineer can score points. Yet, now we are asking design
engineers to share their "work-in-process," even with its incompletenesses.

CONCURRENT ENGINEERING

Concurrent engineering requires all concerned to iterate the development of
the product, process and logistic support designs in a collaborative effort

to find the best possible solutions. Yet, it is easy to fall into the morass of one-upmanship between concurrent engineering team members. Why is this?

Recently, a staff of 20 professional managers all working together on a daily basis, were asked to list the 10 people they knew the best. They were then asked to list next to these 10 names two or three strengths which they admired, or which they felt built upon their work together. They struggled with this request.

Next, they were asked, "If you had to list one another's weaknesses, could you? They all said they could.

If we are blind to each other's strengths, yet are aware of the weaknesses of others, is it hard to understand why we struggle to produce the quality products, processes and services necessary in today's markets? We spend so much time hiding our weaknesses that trust is never built. Moreover, we never give another person's strengths and capabilities an opportunity to flourish because we do not recognize them. In short, we use up so much psychic energy in hiding our weaknesses, that our true capabilities wither.

If we want to get serious about concurrent engineering, we have to face up to some pretty bad habits in our of behavior. Until we learn to appreciate, respect, and build upon our strengths and capabilities, we will never know the power of the synergy of concurrent engineering. Until we learn to "dialogue" with one another, opening ourselves to the inspiration and insights available from others, we won't know what true empowerment is all about. It is ironic, but real empowerment comes when we are no longer concerned about our power, but are able to show genuine excitement over each other's capabilities. This is true dialogue and true knowledge networking.

We also need to understand that there is a difference between sports teams and the teaming process. In sports, the rules are known, the roles and responsibilities assigned, and the task easily definable. In much of our teaming, rules are not known. We make them up as we go. The roles and responsibilities evolve. The tasks are often closely interrelated to other tasks in a larger tapestry of interdependencies. Sports teams play against other teams. In business, we play among and between teams. The exciting thing about teaming is the unique contributions of the individuals are absolutely essential for the success of the group effort. It is not "the individual vs. the group," but both the individual and the team which must succeed in working in parallel through knowledge networking.

KNOWLEDGE NETWORKING

Knowledge networking assumes we have the trust, respect and human abilities to draw upon one another's knowledge, whether we are working in the same room or on distributed teams. Too often, we get bogged down on large cross-functional teams. The dynamics of diversity are difficult for many of us. One solution is to break up into smaller teams and work in clusters of virtual teams. The smaller teams allow for genuine human interaction, and the teamwork of teams in the clusters builds strength upon strength.

We need to realize that these teams are not competing against one another, but working in a larger tapestry of interrelatedness. We need to tap into the wider knowledge resources within our organizations if we hope to achieve both time-to-market and time-to-profitability.

In the book "Fifth Generation Management: Integrating Enterprises Through Human Networking," I have identified some attitudes and assumptions of the industrial era which we need to unlearn: division and subdivision of labor, self-interest, pay for narrowly defined tasks, one-person-one-boss, thinking at the top/doing at the bottom and simplistic automation.

To tap into the knowledge resources in our enterprises, we need to develop an understanding for a new set of principles: peer-to-peer networking, the integrative processes, work as dialogue, human time and timing, and virtual teaming.

The shift from the industrial era to the knowledge era is a shift from working sequentially (and hiding our weaknesses) to working in parallel through multiple virtual teams (and building upon our strengths). It is also a shift from a "command and control" management style to one which "facilitates and coordinates" multiple activities. It is a shift from an environment built on training to one where organizational learning is the norm. Peter Senge's wonderful new book, "The Fifth Discipline: The Art and Practice of the Learning Organization" is a valuable contribution to this understanding. We are just beginning to realize the fruits available through concurrent engineering, made possible by knowledge networking and the learning organization.

Joseph Harrington was wonderfully positive. He knew how to build upon the strengths of others. In his writings, and in his life, he has helped to awaken within us a deeper understanding that life is more than computers and cables: it is people with visions, aspirations, capabilities and knowledge, who can, in dialogue, tap into each other's creativity.

As we move from CIM to CALS to CE and knowledge networking, we will rediscover the value of our technologies, and the value of one another as well. As we begin to value one another, we will be able to add value and quality to our enterprises as well as our everyday lives. "Big" CIM requires people who stand tall together, have vision, build on knowledge and are decisive.

CIM: Guidelines for Implementation

by Thomas J. Billesbach, PhD
University of Nebraska

INTRODUCTION

Struggling with implementation issues can often be a slow and frustrating process. However, without proper planning, an organization can easily find itself automating and integrating a mess. The result is a computerized manufacturing process that can make even more mistakes, but do it quicker. How can a manager minimize the probability of failure? Although there is no "best way," there are guidelines that can increase the probability of success. This article examines those principles and provides a basis for successful implementation.

GUIDELINES

If the Manufacturing Studies Board of the National Research Council is even partially correct, most manufacturing companies will quickly be examining CIM. The board estimates that companies implementing CIM can cut work-in-process inventory by up to 60%, reduce engineering costs between 15-30%, vastly improve engineering productivity and quality, and achieve other significant operational improvements. Until recently, technology was limited, and there existed a genuine lack of understanding of CIM requirements, including equipment and software. That is no longer true. As managers take steps towards the factory of the future, a methodology of implementation becomes ever more valuable.

TOP MANAGEMENT SUPPORT

If you read any article dealing with a significant change in an organization, one of the first things discussed is the importance of top management support. Implementing CIM is no different. In fact, top management support becomes even more crucial to success due to the long duration of the process and the potential capital investment. Re-directing organizational efforts and resources often requires a cultural change. Withstanding pressures to abandon CIM is easier if all members of top management support the effort. Achieving top management's support is easier if they truly understand what must take place for CIM to be successful. Without adequate education and training of top managers, support tends to diminish with time.

SIMPLIFICATION

Richard J. Schonberger emphasizes the importance of simplicity in the manufacturing process (1). Automating and integrating an existing operation without first examining ways to simplify the process is a sure prescription for failure. The elimination of steps within the transformation process makes integration of operations easier. Simplifying an operation makes it easier to automate. A simplified process requires considerably less complex automation which, in turn, makes the integration process easier. The key to success in simplification is through employees. Those individuals currently involved in the manufacturing process are in the best position to identify nonvalue-added steps and to eliminate and simplify the process. Simplification is the foundation for computer automation. Operators and assemblers typically have the greatest insight in eliminating unnecessary steps in the manufacturing process. This, of course, is further facilitated by proper training and education.

EDUCATION/TRAINING AND COMMUNICATION

Employee education and training lays the foundation for the simplification phase. Without proper training and education of employees, supervisors, and managers, changes are often made that are not consistent with CIM objectives. For example, developing a sophisticated conveyor system that stores products between operations results in excess capital investment, long queue times, delayed feedback of product quality, and a more complex system than is needed. To avoid wasted efforts, those involved in the process need to fully understand CIM.

How can a manager get an employee involved in the process when it is likely that person's job will no longer be necessary after the system is fully in place? Very few people wish to be trained and assist with CIM implementation if it results in their job being eliminated. To overcome this, managers need to make the commitment to retrain those employees to perform other duties. An organization faces a dilemma when CIM results in fewer employees, and yet those employees are vital to success. In some companies this is a major concern. However, managers must remember that CIM does not happen overnight. It is typically a long-term process allowing time for natural attrition to occur and retraining to take place. In the event that natural attrition is not sufficient and a limited number of new jobs will be created, managers need to involve and work with a selected number of employees.

Open two-way communication between management and employees is also crucial to implementation success. Too often, operators and assemblers are the last to know, and yet they can provide a tremendous amount of insight. By withholding information from employees, all types of rumors can abound which may slow the implementation process. The extent of sensitive information that must be withheld from employees is considerably less today. How can one expect an operator or assembler to contribute with only half the facts? There is bound to be some apprehension by employees associated with a CIM effort. However, it is easier to manage when the communication channels are kept open.

EFFECTIVE DATABASE AND NETWORK

According to Vaughn Johnson, to move from "islands of automation" to an integrated whole requires local area networks (LANS) and database management systems (2). Up front planning by management in selecting an effective network and database management system is important. Managers must recognize that integration of engineering and manufacturing are important, but so is need to link business functions. Weyerhaeuser managers found a shared database was critical to their CIM success (3). Another organization, Bentley Nevada, found that a unified database that served all functional areas, not just engineering, was a key to success. Without an effective database and networking system, integration cannot be achieved.

IDENTIFICATION OF KEY PRODUCT LINE

Rather than implement CIM on a large scale, it can be very effective to identify a key product line that accounts for a significant portion of revenue and is in its growth stage or early maturity stage of sales. By selecting a key product line, managers, supervisors, and employees can minimize the risk while going through the learning curve. During the implementation of CIM, new and better approaches will become evident. If a large scale implementation process is undertaken, changes, even for the better become more difficult to institute. Small scale CIM projects provide a basis for the organization to adapt and change while reducing the level of risk.

LONG-TERM VISION/PLAN

Because CIM systems take a long time to develop, implement, and refine, a long-term commitment is needed. Managers need to formalize their visionary statement and plan of action. Otherwise, a number of stand-alone applications will result. Although there are benefits to numerically controlled machines, robots, CAD, etc., the greatest benefits are realized when these are integrated into one whole system. A specific and detailed outline of an organization's CIM objectives keeps people and resources focused. Too often, without a well-defined plan, CIM efforts become diluted as employees within the company interpret top management's CIM objectives differently.

CONCLUSION

Will more managers examine CIM in the near future? Is the factory of the future closer at hand than ever before? The answer to both these questions is yes. Will CIM spell success for all those companies that adopt it? CIM is not a panacea for an organization's ills. Implementing CIM for an entire manufacturing process may not be appropriate. What managers must do is examine their operations and identify those areas that stand to benefit the most from CIM. According to a report by Arthur Andersen and Company, "CIM does not produce quick, easy benefits; CIM projects can last several years and cost many millions of dollars (4). CIM is more apt to be a long-term solution.

Following the above guidelines for implementation does not guarantee success but greatly enhances the likelihood. These guidelines are meant to provide some insight to smooth the transition from traditional manufacturing to the factory of the future. They are not a substitute for common sense. Managers of companies that have implemented CIM have approached it in various ways. However, the above-mentioned areas are consistently found in most successful CIM implementation efforts.

REFERENCES

(1) Schonberger, Richard J., World Class Manufacturing, The Free Press, New York, 1986.
(2) Johnson, G. Vaughn, Information Systems - A Strategic Approach, Mountain Top Publishing, Omaha, NE, 1990, p. 440.
(3) Schatz, W., "Making CIM Work," Datamation, December 1, 1988, p. 21.
(4) Arthur Andersen & Company, Trends in Information Technology, 3rd edition, 1987, p. 86.

Strategic Planning for CIM: The Key to Justification and Implementation

by Michael W. Roberts
CAM-I

Lewis J. Soloway
Deloitte & Touche

INTRODUCTION

Increased competition finds more U.S. companies searching for ways to improve their competitive position in manufacturing. Many are coming to realize that the journey to computer-integrated manufacturing is not just a "day trip." In fact, successful companies have realized that achieving the benefits of CIM requires understanding of customer needs, long-range planning, a commitment to learning and most of all an atmosphere of continuous improvement. Essentially, the basis for success in implementing CIM is a sound business strategy that incorporates these four elements. CIM investments get evaluated by how much they contribute to the attainment of strategic objectives.

Through the CAM-I Cost Management System Program, many leading manufacturing companies are conducting research to develop the next generation of decision support tools to aid this strategic CIM investment process. Much has been written about the need for more comprehensive decision support tools to augment the existing financial methods for justifying capital investment in CIM. It has been said that traditional methods, such as discounted cash flow, return on investment and payback analysis often fail to recognize the strategic benefits of investments in factory automation.

In conducting the CAM-I CMS research, it has become clear that many of the key criteria for evaluating capital investments lie outside traditional financial analysis. Making sound investment decisions in computer-integrated manufacturing requires that companies utilize a broader management approach to capital investment decisions which:

* Provides the systematic development of capital investment options through the incremental application of technology.
* Provides the strategic analysis of those investment options.
* Provides the basis for long-range capital planning.

The objective is not to justify investment in automation for the sake of automation. Rather, this strategy identifies the most effective use of a company's investment in capital by developing a strategic plan for long-term manufacturing investment. Strategic planning for CIM must take into consideration the customer's needs, and provide the basis for implementing manufacturing automation as one tool for meeting those needs.

STRATEGIC ANALYSIS OF INVESTMENTS IN CIM

Traditional financial methodologies have often failed to provide the strategic analysis to evaluate and support investments in CIM. This results from the inability of the user to quantify and forecast the strategic benefits of the decision to invest in CIM.

Three aspects of CIM that prove difficult to assess are:

* CIM programs are large, complex, and lengthy.
* CIM benefits usually increase with time.
* CIM benefits expand due to integration synergy.

These benefits are often described as the "intangible" or "nonfinancial" benefits of CIM investment. In fact, the benefits are real and over the long-term, are financial in nature. For example: a CIM investment that improves product availability and customer delivery can be expected to increase sales and market share over a period of time. The difficulty in analyzing these benefits is predicting the amount of increased sales and when those sales will occur. In many cases, financial analysis completely ignores these benefits rather than rely on estimates of future events that are difficult to quantify. As author Bob Kaplan observed, there seems to be a preference to being "precisely wrong, rather than approximately right."

A second problem is that the financial analyses assume the "status quo" as a basis of comparison for the CIM investment. This comparison biases the decision process. It ignores the probability that the competition will continue to improve its manufacturing capabilities through CIM and other methodologies. The status quo rarely lasts. It is replaced, at the least, by a projected continuing deterioration in competitive position including lower market share, higher operating costs, etc.

In the CAM-I CMS program, we refer to this as "capital decay." Capital decay occurs when improved technologies or other competitive pressures erode the value of the producing capability of existing manufacturing equipment or processes. As with the strategic benefits of CIM, this erosion of value is difficult to quantify, and often ignored. One solution lies in the process of setting investment criteria.

There are three elements to the establishment of investment criteria. The first is to identify strategy and tactical goals. These include business objectives, product forecasts, competitive functional strategies, and organizational goals.

The second element is to translate the goals and objectives into critical success factors and performance targets. These targets include financial and nonfinancial quantitative targets and qualitative targets. These targets have to be considered within the available resource capacity and constraints. Examples of financial targets are ROI and NPV. Nonfinancial quantitative target examples include throughput time, process yield, schedule attainment, and leadtime reduction. One way to think about qualitative criteria is that these should represent nonquantifiable strategic considerations, such as the degree of manufacturing flexibility achieved, the degree of process innovation achieved, the degree of satisfying customer needs, etc.

The third element of establishing investment criteria is to identify value-added and nonvalue-added costs and their cost drivers. This would be accomplished through activity analysis, value analysis and target costing. It is through the analysis of the cost drivers and causes that we can establish performance targets for those items not stipulated in the strategy-to-critical-success factor translation.

IDENTIFYING CIM INVESTMENT OPTIONS

There is another benefit of identifying the generators and causes of value-added and nonvalue-added costs. The analysis can uncover additional opportunities for capital investment. These opportunities support the continuous improvement process.

Today, most investment decision approaches focus on the costs incurred rather than the sources of cost generation. They focus on specific organizational cost controls rather than upon the interorganizational cost performance dependence. A cost control emphasis will not, in many instances, help eliminate or reduce the nonvalue-added elements of a company's costs because cost control narrowly ignores the cross-functional generators of cost. Activity analysis is the means to eliminating or reducing nonvalue-added costs by focusing on and assigning costs to specific activities within organizational units and then focusing on the generators of these costs.

Conducting this analysis is a crucial first step in developing solutions to three fundamental questions every world-class manufacturer must answer:

* What are the key activities and processes of the enterprise, and are they efficient?	COST
* What are the causes of rework, scrap, or other quality problems?	QUALITY
* What current operations constrain product throughput?	TIME

Some of the data required for analyzing cost and quality issues may be found in the existing cost accounting system. Often, this can be used as a starting point for understanding the overall magnitude of manufacturing costs. Additional information may be found by reviewing manufacturing routings, procedures, and job descriptions.

It is unlikely, however, that the information in most current systems will be sufficient to determine or evaluate the causes (drivers) of cost, quality, or throughput problems. In order to fully understand these issues it is necessary to understand the cross-functional relationships which are often an integral part of the manufacturing process. In many cases, the use of flow charts or other modeling techniques may be required to provide additional understanding as to the manufacturing costs and their drivers.

Companies implementing activity based cost management systems based on activity analysis have found these systems provide valuable insight into understanding the costs and causes of manufacturing activities. This enables managers to streamline operations by eliminating nonvalue-adding activities. It also assists in pointing out opportunities for automation of manufacturing processes. Companies that have successfully implemented CIM have realized the need for well-documented manufacturing processes and supporting procedures. The necessity for documenting the manufacturing process increases with the level of computer-integrated manufacturing.

IMPROVE BUSINESS PRACTICES FIRST

Understanding the activities within a manufacturing process, makes it apparent that some activities can be simplified or eliminated without changing the product design or introducing automation. Utilizing JIT, TQM and other world-class manufacturing philosophies can eliminate waste and inefficiencies with little financial investment. Many firms have made substantial investments in CIM to automate nonessential functions and activities. The old systems adage "identify, simplify, automate, integrate" is sound advice for implementing CIM. There is nothing more wasteful than automating an unnecessary function. Eliminating and simplifying manufacturing activities reduces cost, as well as the number of processes to be automated and makes better use of funds devoted to automation. Activity analysis can pay for itself and the initial investment in CIM projects by first simplifying or eliminating the nonvalue-added activities and their drivers.

Having identified and simplified existing manufacturing processes, the next logical step is to prioritize the opportunities for automation. Companies vary in the way they approach the process of prioritizing investments in CIM. One of the best approaches is competitive benchmarking. Benchmarking is the process of comparing a company's existing manufacturing processes against those of a competitor's best internal practices or new technologies. This comparison should take into account anticipated impacts of CIM investment on the cost of production, improvements in quality, improvements in customer service, and delivery time.

One of the best sources of information for benchmarking is a company's customers. The customer often can pinpoint opportunities to improve quality and delivery and provide insight for future product requirements. Another excellent source is a company's suppliers. Suppliers can provide information about the anticipated availability of new technology and industry trends and competition. Companies with multiple divisions should utilize internal comparisons of best manufacturing practices between divisions. Other information sources include trade publications and trade shows.

These efforts will result in identifying a series of investment options. They can then be evaluated from the perspective of four strategic requirements: customer needs, strategic enhancement, core competencies, and continuous improvement.

DECISION SUPPORT TOOLS FOR EVALUATION

Once the investment criteria have been established and opportunities identified, the next step is to evaluate alternatives and select opportunities. Begin by evaluating competing project alternatives. This needs to be done with a consistent approach. The financial and nonfinancial quantitative and qualitative factors must be considered and risk must be assessed. A possible tool to this approach is a multiattribute decision model (MADM).

MADM is a tool to help managers make cost-effective investment decisions. This model scores investment opportunities according to their strategic significance. It evaluates the potential impact of the investment on the overall goals and objectives of the firm. The model provides a weighted ranking by reviewing critical success factors against policy input. Each of the factors is weighted based on relative importance. Values are assigned for different performance levels of the factors. Risk is presented as a level of confidence that each factor's forecasted performance is attainable. The candidate project's score is calculated by multiplying each factor's weight by its performance target value and its risk level, then summing each factor's multiplied totals. The score can be used to rank investments in an order of relative priority.

MADM type models have been developed by the National Institute of Standards and Technology (NIST). They can also be developed on spreadsheets. They are relatively easy to apply once strategic weights and decision tables are established.

The second step in the evaluation process is to select an investment portfolio. This determines the portfolio of projects that best meets our strategic objectives. Projects cannot be evaluated in a vacuum. They must be considered in terms of other projects both in-process, and planned. The total costs and benefits must be defined and the risks must be assessed for the entire portfolio, as well as for the individual projects. This is where the synergy we alluded to previously must be derived.

Resource constraints must also be considered. We have only limited amounts of resources (capital, people, competencies, etc.) to invest in projects. Knowing this, we can reduce the overall project portfolio on one hand or, as an option, look for avenues to increase the needed resources since they have now been identified.

Specific means to evaluate investment portfolios as a whole are only in their infancy. Both cross-sectional (functional) interdependencies and cross-temporal (time-phased) interdependencies need to be evaluated. Jim Reeve of the University of Tennessee and Bill Sullivan of Virginia Polytechnic Institute have studied this issue for the CAM-I CMS Program. They suggest that simulation coupled with decision flow networks (Monte Carlo analysis) may offer the best approach to evaluating the portfolio of investments. Reeve and Sullivan have also studied the combination of simulation, Monte Carlo analysis, and MADM with some promising results.

SUMMARY

We have looked at the process, tools and implementation of an investment justification methodology for CIM technologies. In closing, there are several things to consider:

* Short-term goals must be tied to the strategic goals. We must take the broad perspective.

* Explicit and consistent evaluations must occur. We evaluate competing technologies in the same manner with the same criteria, the same weights, and ranges of values. We evaluate the portfolio on the same basis. It is only when the strategic objectives change that we reassess the projects in the portfolio under new criteria or new weighting or new values for performance.
* This is a consensus-based evaluation and selection process. The management team must know, understand, and believe in this process, and be involved in it. It cannot be done without their knowledge or acceptance.
* Qualitative and nonfinancial quantitative critical success factors must be considered in the evaluation process. These attributes are just as important, if not more so, than the financial criteria. They provide a means to translate the strategic goals and objectives into the decision-making process.
* Risk assessment is a must. While there is considerable argument as to how risk is incorporated in the evaluation model, it must be incorporated whether directly in the model or separate from it.
* Performance tracking and post-implementation audits are essential. We never know how well we are doing in our decision-making process unless we go back and review. We never know how well the project is doing unless we have project milestone reviews. If there are variances, we can adjust to take advantage of missed opportunities, or move to mitigate risk.
* This investment management methodogy allows us to consider the iterrelation of the many projects that occur in the consideration of an integrated strategy and CIM. The portfolio analysis approach is our means of tracking these projects.

The lack of general acceptance for these approaches to capital investment justification is largely cultural. The capital budgeting processes and traditional financial evaluation methods are well ingrained in corporate culture and will take years to change. It is imperative that organizations begin to address the strategic issues involved in CIM justification. Interim measures could include adding a written assessment of strategic impact as part of the standard "capital authorization request."

REFERENCES

1. Berliner, Callie and Brimson, James, A., Cost Management for Today's Advanced Manufacturing. Boston: Harvard Business School Press, 1988.

2. Brimson, James, A., Activity Based Investment Management. AMA Membership Briefing. New York: American Management Association Publishing, 1989.

3. Hayes, Robert, H., Wheelwright, Steven, C., and Clark, Kim, B., Dynamic Manufacturing. New York: Free Press, 1988.

4. Johnson, H. Thomas, and Kaplan, Robert, S., Relevance Lost: The Rise and Fall of Management Accounting. Boston: Harvard Business School Press, 1987.

5. McNair, Carol, J., Norris, Thomas, F., and Mosconi, William, Beyond the Bottom Line: Measuring World-Class Performance. Homewood, Illinois: Dow Jones Irwin, 1989.

Part 2

Tools and Technology of CIM

Mastering Work with EventMap™ Diagramming: A Tool for Developing Your CIM System

by James B. Johnson
Berkeley, CA

INTRODUCTION

The purpose of this article is to provide education and issue a challenge about a new approach to improved organizational performance. It is based on a powerful new visual process modeling skill, called EventMap (TM) Diagramming. This method has been used in the design and implementation of a very successful CIM system enhancement.

People have to understand a work process before they can effectively perform it. However, it takes even greater understanding to improve a work process. For a simple one-person job, such understanding can be achieved by practicing and experimenting with each work step. But for an entire work operation involving more than a few people, that is seldom feasible. People achieve their understanding of complex work processes and how to improve them by creating models of them in thought, on paper, and with groups.

This is difficult and time consuming and often ends without complete understanding or agreement. Yet, it is very important. All performance measures including cost, quality, productivity, flexibility, cycle time, time to market, customer service and job satisfaction, are determined primarily by the individual and collective understandings and agreements about the work process. These understandings and agreements are the result of specific activities, from policy making to ad hoc procedure changes. How can these activities be made more effective and efficient to produce better understanding and agreement more quickly?

CASE EXAMPLE

Dave is the owner and manager of XYZ Company, a medium-sized machine shop, producing precision parts to order for local manufacturing firms. Business is good. Dave keeps the shop running around the clock. But now his largest customer tells him that he needs to have his orders filled within two days, instead of the usual three, or he is going to have to pull the business in-house. The customer is willing to pay extra, but Dave knows the premium will not cover the cost of more floor space, equipment and people. Dave is going to have to get that extra 24-hours out of his existing operation if he wants to keep that business.

FIRST APPROACH

Dave has called the shift supervisors, Bill, Tony, and Mike, and the lead operators, Tom, Joe, and Larry, into a planning session to see what can be done. Sitting around a table in the lunch room, they try to tackle the problem. There is plenty of talk, and they use the white board to review the latest production figures. Several hours pass. Many ideas are thrown out, but little progress is made. At the end of the day, the only thing Dave knows for sure is that he has lost several hours of production, and that he does not have an answer for his customer. And having no answer is as good as having the wrong answer.

Dave does not know it, but his problem has to do with vision. In this case, it means literally what he and his employees can or cannot see. The shift supervisors can see the operations areas filled with people and equipment, parts and packages. Some people are at the job, others are taking breaks, some work is going smoothly, some things need fixing. The lead operators can see their work stations, the equipment, their hands in motion, the metal blanks being cut and drilled. Sometimes the work flows smoothly, sometimes things go wrong. Everything seems busy all the time. There's no extra time for anything.

Dave can see all this too. He can see that everyone is working to full capacity. There are no major personnel problems. Dave can visualize each step of the entire operation, one step at a time. He has got the production figures completely memorized. And he can swap descriptions of what he sees with the men in the planning session right up until the deadline for getting back to his customer.

THE EventMap (TM) APPROACH

What Dave cannot see, and what he needs to see, is his entire operation in a single, integrated view, one that he can share with his employees, and one in which he can find that 24-hours. Something like the diagram in Figure 1.

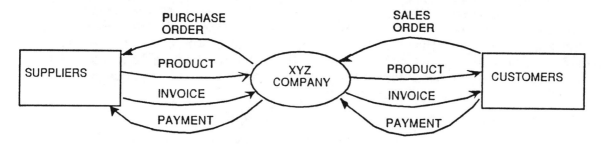

Figure 1. XYZ Company main operations overview.

Not much detail and no real surprises and yet, looking at the diagram in Figure 1, Dave finds it a little easier to focus on the problem. He thinks about the fact that the "purchase orders" for the customer in question are frequently pre-worked parts that the customer has already ordered from the supplier and delivered by the supplier to Dave for finishing. And he notices that this problem does not have anything to do with the ordering or billing cycle; it is just getting the machine work done faster. "What am I going to have to do?" Dave thinks, "Give up my smaller customers to hold on to this big one? If only I could see the problem more clearly."

Now, suppose that Dave could see the problem more clearly. Suppose Dave could see the next level of detail in his operation in this same diagram format, as shown in Figure 2. Better yet, suppose Dave knew how to draw diagrams like this. And even better still, suppose Dave could draw a diagram like this while his employees talked about the work process during the planning meeting, getting the facts directly from the people with the greatest knowledge of the work, and keeping the discussion organized. And when the diagram was complete, everyone agreed on all points, producing a detailed agreement on the entire diagram in one pass.

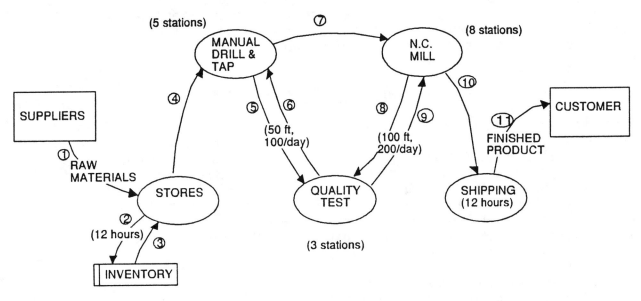

Figure 2. XYZ Company major customer machine work.

In order to do this, Dave follows a simple set of rules. He draws ellipses for activities he can directly control and rectangles for activities which he cannot. He draws thin rectangles with a double left sides for storage locations or containers, and labels all these parts with easily understood names and measures. Then he traces the movement of parts and information among these activities and locations using arrows.

He makes sure that the arrows stand for real things really moving from one place to another and does not use them for "do this next", or "wouldn't it be great if", or "these are somehow related". That way, all parts of the diagram are connected by cause and effect logic and can be directly verified and measured, just like the real work process. And all he has to do to fill in the arrows is ask, "What else happens?" and "What happens next?" He gives names to some of the arrows, but only sequence numbers to others, since it is clear what they represent.

When the shift supervisors look at the arrow from "Stores" to "Manual Drill and Tap", they easily visualize someone carrying a box of parts from the store room out to the operations area. When the lead operators look at the ellipse labeled "NC Mill", they easily imagine themselves standing in front of the big machines. When they talk about those things, they point at those figures on the diagram. Dave, standing up by the white board, points to them also, so everyone knows exactly what's being discussed right now. If

Dave can't figure out what to point at, he knows he has to either add something to the diagram, or get the discussion back to what really happens.

As a group, the planning team easily makes the decision to just diagram the main movement of materials and leave the paperwork out for the now. Likewise, they make tacit agreements about what they do not bother to put on the map because it is so well known by all, such as the fact that the "Stores" and "Shipping" functions are done by the same people who run the work stations, on a rotational basis.

Looking at the diagram, they are able to shift their attention from the output product numbers, to how those numbers are produced. They are able to put what they know from common sense experience into an integrated picture of how all the pieces add up. Suddenly, they start seeing that maybe they can speed up the process, without burning everyone out.

EventMap (TM) DIAGRAMMING IN ACTION

"Why should the Quality Test station be twice as far from the NC Milling station as the Manual Drill and Tap station is, if NC Milling requires twice as many quality tests?" "Well, that is the way it was set up originally before we got the extra assembly area and moved the NC Milling operation into it." "I guess we were just so busy, we didn't notice how much further we were walking. Maybe the guys liked the break it gave them." "Yeah, but how much time does it take to walk that far each day? Let's see, that's, 100 times 200, that's 20,000 feet. That's FOUR MILES!! That's got to take an hour each day right there!"

"Except that's only one way! And it does not account for the time the operators spend waiting to pick up the last test job they dropped off." "So you mean that cycle could be taking up three hours a day? If we could knock that off, that would mean there's only 21 to go." "Hold on. That's three hours of production time, not order cycle time." "Okay, but now at least we are getting somewhere. What else is there? What's this 12 hours that it takes to get something put on the shelf in inventory?" "That's just what we figure the average is. The stuff comes in and could sit there through two shifts until someone gets around to it. Depends on the backlog. Sometimes stuff sits out in the receiving area for two or three days." "WHAT!?!"

Using this type of diagram, the group discussion stops diverging and starts converging. The diagram is so effective at describing the work process so quickly, that people start to talk about the work process as if it were right there up on the board. Everyone knows what all the parts mean. They either happen in the real work or they do not. If there is a disagreement, you can get more information, or simply observe the process in action. The diagram is not carved in stone; when there is an error on it, you just scratch it out and make the correction. In fact, the diagram is so flexible and so accurate at describing the work process, that people quickly start expressing their ideas about change directly in terms of the diagram.

"You want crazy ideas? You got 'em. Suppose that arrow went directly from "Inventory" to "Manual Drill and Tap." "How's it going to do that? You can't just get rid of the "Stores" function?" "Why not? It's all just work steps isn't it? We've got everybody doing a piece of every job as it is.

Why not scatter it out a little more?" "I can't tell if you're just being
sarcastic or what. What do you think, we can just have the suppliers dump
the stuff on the receiving dock and fish it out from there?" "I don't know.
Half the time the operators are doing that anyway. It takes so long to get
stuff on the shelves."

"Wait a minute, maybe that's not such a crazy idea after all. Let's draw it
like that and see what it looks like." "Hey, while you're at it, do the
same thing to the shipping function. I don't know what it means in terms of
real work, but it'll look the same on the diagram as what you are doing to
Stores." "Yeah, and while you're doing that, move "Quality Test" so it's
not so far from the other work stations."

Within a few minutes, the planning team looks at the diagram in Figure 3.

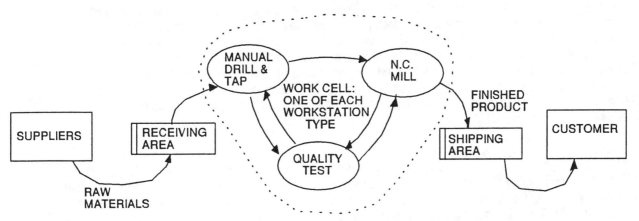

Figure 3. A diagram of the new operations of XYZ Company.

"You don't have the same number of work stations of each type to put them
all in work cells." "We don't have to: we just have to set up two work
cells for our major customer and get some more test equipment, which is the
least expensive." "What's this about the finished product going magically
to the customer?" "Some of the mill operators are doing the shipping now on
rotation; in the work cell, they'll just do it as they go along. All they
have to do is drop the finished part in a slot in a box and seal the box
when it is full. Break up the work. They can choose how. The labels can
all be made in advance." "What about the freight company?" "Maybe the
customer will take that on; they want the service, don't they? It's a lot
less hassle than pulling the whole operation in-house."

"And people are going to just paw through piles of stuff on the receiving
dock?" "Well, the security back there is pretty good". "We can set up some
well-labeled shelving out there. And we've got a choice of suppliers. One
of them will probably be willing to put the boxes in the right place. The
customer gets a say there too."

"Is this really going to save enough time?" "I don't know, but we can trace
it out on the diagram." "You know, maybe we could draw diagrams for the
inside of each of the work station activities, show them to the operators,
and ask them if they think it would work". "Next thing you know it's them
that's going to be telling us how to make it work."   "Do you think we can
really pull this off?" "I don't know, but it's probably a lot easier to do
this than it is to replace our biggest customer. What do you say we give it
a try?"

CASE SUMMARY

The case of XYZ Company may be somewhat oversimplified. Few groups can take on such changes "within a few minutes". But when people can see where they are going, they are not nearly as afraid of getting started.

Dave and his group were fortunate. They had the use of powerful new tools for describing work, called EventMap (TM) Diagrams. EventMap (TM) Diagrams do in work what road maps do in travel. They help us master the activity by giving us perspectives on our surroundings and guidance toward our goals.

THE TOOL, THE SKILL AND THE PRACTICE

EventMap (TM) Diagramming is a set of use procedures and graphic conventions added to an existing diagram format, called data flow diagrams. Data flow diagrams are the central tool of structured analysis, a well-defined, widely used, and openly available systems engineering methodology. The diagram format, while consisting of only circles, arrows and boxes, is based on a mathematical system for representing cause and effect logic and is different in structure and utilization from any other type of visual illustration method.

EventMap (TM) Diagramming can be performed on an individual basis, with one's work group or as a facilitation service for other groups. As a tool, EventMap (TM) Diagramming requires only pencil and paper or flip charts and marking pens. As a skill, it can be grasped within a few hours. As a practice, it can be mastered within days.

Review the diagram in Figure 4 and ask yourself what you know about how the activities depicted in the diagram are performed at your company or at a company you are familiar with.

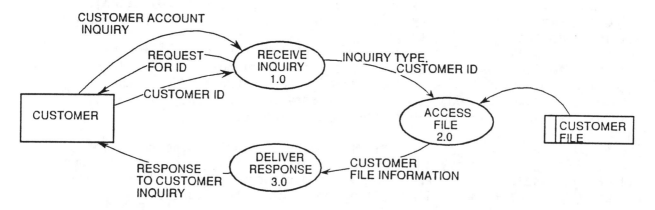

Figure 4. Customer account inquiry and response.

Accurately labeled circles, arrows and boxes of an EventMap (TM)) Diagram represent both the action steps of work and the movements of work products and resources. Review how this is represented in Figure 4. Imagine these graphic parts being divided and re-divided to precisely follow the flow of activity in your real version of this work process until you are looking at steps and movements which can be performed immediately and measured to the penny and the second. EventMap (TM)) Diagrams possess common sense visual qualities which allow you to recognize where parts may be missing or out of place as the diagrams are assembled. These same qualities allow you to quickly locate bottlenecks, shortcuts and redundancies, and many other features of the work being examined.

Starting from a detailed level, EventMap (TM) Diagram parts can be re-combined to accurately summarize large-scale resources and results. The diagrams can depict overviews of corporate divisions and markets, details of account transactions and desktops and all the connecting levels in between. These work structures may be mapped from different points of view, including those of executive managers, special departments, individual employees and outside stakeholders. Consider all the activity in Figure 4 as a single process ellipse. Ask yourself what else goes into and out of this overall activity and what other processes would be connected to it through those transactions.

EventMap (TM) Diagrams operate in conjunction with traditional descriptive tools such as speech, writing and illustration, math, outlines and tables, and catalogs, spreadsheets and schedules. Review where these tools might be used in your example of the activity in Figure 4. EventMap(TM) Diagramming can serve as a powerful information gathering method for these other tools, or can re-express the information from these forms for more direct visualization, discussion, and action.

Likewise, as a basic descriptive tool, EventMap (TM) Diagramming can accelerate progress in existing methodologies and improvement programs. Change ideas recorded during the modeling activity feed directly into structured project plans. Consider what you could recommend as changes to your example of the activity in Figure 4. Which of those changes would appear as a verifiable change to a specific part on the diagram?

UNIVERSAL WORK PROCESS LANGUAGE

With the EventMap (TM) use procedures and graphic conventions added, this highly accurate, powerfully analytic diagram format can be used in fast-moving, fully interactive group sessions. Participants require virtually no prior knowledge of the method to use is. They can range in position from factory workers and secretaries to staff professionals and senior managers. Session purposes can include managing change, building consensus, solving problems and organizing production. Session content can range from administrative paperwork to factory floor work station layouts, and from preliminary executive consensus to detailed technical simulations.

The skill has been successfully used across these ranges of participants, purposes and contents in dozens of organizations including Hewlett-Packard, Chevron, National Semiconductor and Pacific Gas and Electric Company.

EventMap (TM) DIAGRAMMING IN THE DEVELOPMENT CYCLE

In a typical technology implementation cycle, as shown in Figure 5, EventMap (TM) Diagramming can first be used to capture a new idea for the work process and help translate that into an understandable proposal. Next, in an executive planning session, the diagrams can help achieve agreement on the overall objectives of the program. The skill can then be used to very rapidly create an accurate understanding and solid consensus on the current organizational process and the improvement opportunities within it.

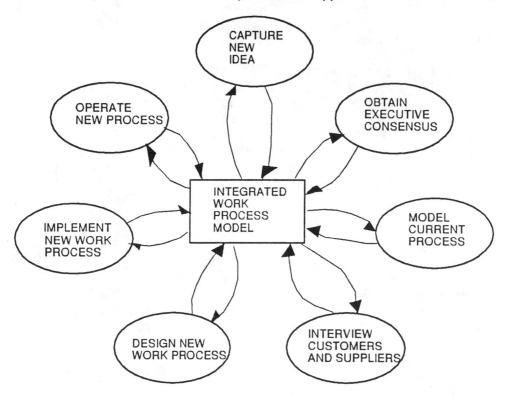

Figure 5.   EventMap (TM) diagramming in the development cycle.

Diagramming sessions can be conducted with customers, suppliers and other business partners to determine how their specific processes will be addressed in the development effort. EventMap (TM) Diagramming can next be applied to design and test out with the full participation of the people who will do the work, the complete new work process from administrative paperwork to technical equipment operations, even designs for computer software. Another set of diagrams can be created to plan the detailed implementation of the new work process design. Finally, EventMap (TM) Diagrams can be used as blueprints for ongoing control of the operation, training of new personnel and continuous improvement of the new work process.

What tools do you and your organization currently use to achieve understanding and agreement in each of the activities in Figure 5 ? To what extent are these tools integrated from one development phase and content area to the next? To what extent do these tools operate on and from an integrated information base?

CONCLUSION

In every application, EventMap (TM) Diagramming produces a significant benefit: its use greatly reduces the amount of time required to achieve a solid agreement based on real understanding. The right decisions are made more quickly. Results increase in both production and performance improvement. This benefit can be directly realized in virtually any organizational performance area including quality, productivity, cost effectiveness, cycle time, profitability, customer service, and job satisfaction.

EventMap (TM) Diagramming has only one significant drawback. There is no theory of operation or conceptual underpinning which contributes in any way to the results offered in this approach. The benefits presented in this article are strictly related to the actual performance of the skill in real work situations. Acquiring, distributing and using the skill are essentially all that matters. You have to actually do it.

Descriptive capabilities are the keys to understanding, communicating and controlling work. These results are in turn the keys to managing change, developing consensus, solving problems and organizing production in work. EventMap (TM) Diagramming gives you new levels of descriptive capability as the key to mastering work.

The skill is easy enough to learn. People have begun using it after seeing it presented in demonstration sessions. The rest follows from practice and common sense. This article actually contains all the information a person needs to get started. Do you have a pencil and paper handy?

Communications for CIM:
An Insight into European Activities

by S.C.H. Withnell
British Aerospace

INTRODUCTION

The Commission of the European Communities (CEC) has established a work program known as ESPRIT, the European Strategic Program for Research and Development in Information Technology.

ESPRIT is a framework which supports many individual projects related to the Information Technology industry. Within ESPRIT, there are four work areas: microelectronics, information processing systems, office and business systems, and computer-integrated manufacture (CIM).

Each ESPRIT project meets set conditions in terms of structure and objectives. The project team must consist of at least two industrial partners from at least two member states. Within the project a mix of users and vendors is encouraged, as is the participation of academic institutions and research establishments.

Each ESPRIT project receives up to half its project costs as a grant from the European Community. To put the scale of the ESPRIT program into perspective, total European Community funds authorized for ESPRIT now exceed 2350 million ECUs. (The European Currency Unit is approximately 1.25 U.S. dollars.) This corresponds to a gross program budget of some 4.7 billion ECUs.

ESPRIT participants retain intellectual property rights to project results. However, there are certain obligations to disseminate general project results into the public domain. While ESPRIT projects are concerned with pre-competitive research and development, early exploitation of results is encouraged.

By examining the ESPRIT-CIM work area, it is possible to determine how the ESPRIT framework relates to the manufacturing sector. ESPRIT-CIM objectives are:

* Strengthen the capabilities of indigenous community CIM vendors.

* Improve the competitiveness of European Community manufacturing industry.

Under Phase I of the ESPRIT program 32 CIM projects were initiated, supported by 150 participants comprising:

* Large industrial companies.

* Small and medium sized enterprises.

* Research institutions.

The ESPRIT-CIM work area is divided into a number of topics:

* Manufacturing systems design and implementation.

* Product design and analysis systems.

* Robotics and shop floor systems.

* Management and control of manufacturing processes.

The fifth CIM topic is CIM architecture and communications which provides a common backbone to these application areas.

ARCHITECTURE AND COMMUNICATIONS

Companies with existing investments in computer assisted design and manufacture are frequently frustrated by the incompatibilities of languages, data formats and communication protocols. It has been stated that between 50% and 70% of CIM integration costs are a direct result of these incompatibilities. This greatly inhibits progress, especially in small and medium sized enterprises.

Three CIM projects are particularly aimed at areas where success would result in a significant breakthrough for the project teams and the ESPRIT community.

ESPRIT project 688, known as AMICE (a European Computer-Integrated Manufacturing Architecture) is developing a generic architecture for CIM. This work is based on open systems concepts and is known as CIM-OSA. Project results are fed from European representatives into ISO TC 184. The project consortium is comprised of 21 users and vendors of CIM oriented systems.

ESPRIT project 2617, known as CNMA, addresses a communications network for manufacturing applications. CNMA objectives are promotion, implementation, and validation of international standards in the area of manufacturing communications. As such, the project is closely associated with the MAP and TOP initiatives in the United States, and has had significant influence on the development of international communications standards. This is particularly true of IS 9506, the Manufacturing Message Specification (MMS), a major service in MAP version 3.0.

ESPRIT project 623 (operational control for robot system integration into CIM) provides methods and tools for the functional integration of robot systems into the CIM environment. Subsystems for explicit robot programming, the planning of assembly operations, and optimizing production layouts have been developed, and are in use. In line with ESPRIT objectives, consortium members have played a leading role in the development of the IRDATA robot programming language which is an accepted work item in ISO TC 184.

To complement the development program, ESPRIT-CIM has initiated an infra-structure activity known as "CIM-Europe." The objective of CIM-Europe is to consolidate and enhance the effects of CIM projects by fostering interaction between ESPRIT projects and other workers in the field. CIM-Europe is based on eight special interest groups which are aligned with topics related to CIM. The major activities of CIM-Europe are an annual conference and technical workshops.

To complete the discussion of the ESPRIT program, it is necessary to describe how the overall work plan is maintained, and how companies can become a part of the program. Each year the ESPRIT work plan is updated through a consensus process. After approval by the European Council of Ministers, the plan is published in the official journal. Associated with the publication of the work plan, is the call for proposals, or CFP. The call for proposals outlines the types of projects sought with regard both to the work plan and the availability of funds.

Evaluation of proposals from prospective participants is managed by the commission, using teams of qualified external evaluators, under strict confidentiality. Assessment criteria includes technical and strategic value of the proposal, the ability of the proposers to complete the work and the "fit" of the proposal with the ESPRIT work plan. Historically, proposals received are more than five times the number that could be supported.

Potential project participants are assisted in finding suitable project partners. Applicants indicating their capabilities and areas of interest are matched with other applicants with complimentary capabilities.

COMMUNICATIONS NETWORK FOR MANUFACTURING APPLICATIONS

Factory automation is characterized by the large variety of computers and controllers from many vendors in any real installation. Accordingly, installations often appear as autonomous "islands of automation."

The user is faced with two options for integrating facilities: one is to purchase computers and controllers from a single vendor and use that vendor's proprietary protocols to achieve communication between the different devices. The user is then tied to the products of that vendor and may be restricted to a limited choice of devices. The alternative is to integrate computers and controllers from more than one vendor, but at considerable expenditure, to achieve communication between numerous proprietary systems.

It is worth reviewing the benefits of successfully integrating manufacturing facilities.

    * Improved quality
    * Accurate and timely feedback of production data
    * Reduced response times to product changes

The bottom line is reduced costs. All these elements lead to a solid improvement in the competitive position of a manufacturer in the market place. An obstacle in the way of achieving CIM has been the high cost of integrating heterogeneous CIM systems. To provide a cost-effective solution, a single communication system supported by many vendors is essential. The evolution toward computer integrated-manufacturing must ensure that common standards are adopted by both vendors and users, especially in the key areas of Industrial Local Area Network communications and data management.

For some years, the Open Systems Interconnection or OSI has been proclaimed as the "enabling technology for CIM." Work in this area has been carried out under the auspices of the International Standards Organization's seven-layer model for Open System Interconnection.

The most publicized examples of this work has been MAP (Manufacturing Automation Protocols lead by General Motors) and TOP (Technical and Office Protocols lead by Boeing) in the United States. This work has now been taken up by the standards bodies such as the National Institute for Science and Technology and the Corporation for Open Systems.

It was seen as vital for European industry that there should be a European program to compliment the other international initiatives in the field of manufacturing networks. This led to the formation, in January 1986, of a project addressing a Communications Network from Manufacturing Applications, known as CNMA, and supported by the Commission of the European Communities under the ESPRIT initiative.

British Aerospace is leading the powerful CNMA consortium to which many major European organizations have contributed. Since the inception of the project, the following organizations have made significant contribution to the work:

USERS

Aeritalia
Aerospatiale
British Aerospace
BMW
Magneti Marelli
PSA Peugeot
Renault

VENDORS

Bull
GEC
ICL
Nixdorf
Olivetti
Robotiker
Siemens

RESEARCH INSTITUTES

Fraunhofer - IITB
University of Porto
University of Stuttgart - ISW

SYSTEMS HOUSES

Alcatel TITN
Comconsult

A specific aspect of the project has been to provide a conformance testing framework for validation of vendor's implementations. This has been achieved with the support of some additional organizations, including SPAG, ACERLI and the Networking Center.

CNMA's current requirements for conformance and interoperability test tools have been met by a complimentary ESPRIT project known as TT-CNMA, lead by SPAG.

The user companies have great experience in computer-integrated manufacturing and have brought to the CNMA project a clear understanding of the user's requirements, and the limitations of current technology. The vendors recognize CIM as a major market area for their products, which include computers and the full range of controllers which modern automation demands.

It is important to recognize that this European wide combination of experience skills, and interests, all dedicated to a common goal, has been the major contributing factor towards the success of the CNMA program.

Clearly, the work of the CNMA project has been a major undertaking. The commitment of the CNMA partners is based upon a firm belief in Open Systems communications as the enabling technology for CIM.

It must be stressed that CNMA is targeting the same final profile for manufacturing applications as the MAP program. However, MAP represents North American needs. In order to provide balance, CNMA acts as a complimentary program representing the European user's requirement for multivendor interworking down to the level of device controllers. Within a manufacturing cell, large American industries use primarily a single vendor's controllers and proprietary communication systems. (Their main need is for a means of linking together a variety of different manufacturing cells.) However, many companies do not have the same level of investment, and demand multivendor working right down to the level of robots, machine tools, etc. Whereas file transfer can solve many problems at the factory backbone level, automation protocols are essential if open systems are to be achieved at this lower level.

At the outset of the project, the following objectives were established:

* specification, implementation, validation, demonstration and promotion of standards and specifications for factory automation applications to ensure the development of standards suitable for European users.

* compatibility with:

  MAP and TOP specifications, and CEN/CENELEC supported standards, in Europe, to insure that a single international profile is obtained.

* promotion of European acceptance of standards to encourage European vendors to adopt them.

* encouraging the creation of validation centers to facilitate testing vendor implementation.

To achieve these objectives, the CNMA consortium engaged in the following main activities:

* specification of CIM user's requirements;

* selection of an unambiguous profile of communications standards to meet the user's requirements;

* implementation of the profile on controllers and minicomputers;

* development of conformance test tools and conformance testing of implementations;

* validation and demonstration of the implementation in real production facilities;

* promotion of the profile.

Considerable effort has been expended to establish standards for industrial LANs, however, there is insufficient information on the needs these standards are intended to satisfy. CNMA has enjoyed the advantage of including seven major industrial users in the project and has conducted detailed analyses of the subject. These cover a wide variety of requirements in batch manufacturing and the process control industry, including costs, time constraints, reconfiguration, redundancy, reliability, integrity and training.

The first activity in meeting the user's requirement is the selection of a suitable, unambiguous profile of communications standards, incorporating both existing and emerging standards. This work complimented other activities leading to version 3.0 of the MAP and TOP specifications. The profile is documented in an Implementation Guide which becomes the specification for all CNMA implementations.

To prove each implementation conforms to specified standards, it is necessary to subject it to "conformance tests." The Fraunhofer Institute, the project's independent test organization obtained and developed conformance testing tools and testing procedures to verify the CNMA implementations in the first phase of the project. The institute worked with The Networking Center, SPAG Services and ACERLI to develop testing tools and procedures for phases 2 and 3 of the project. In the current work, Fraunhofer - IITB participates in both CNMA and TT-CNMA, providing a center of excellence for conformance testing issues.

Perhaps the best validation of the CNMA implementations is by applying them to real production machinery provided by the users. These "pilot facilities" also ensure that the standards are appropriate to industrial use and provide the opportunity to demonstrate the project's achievements.

Finally, the CNMA consortium promotes the communications profiles to encourage their widespread acceptance. This is vital to achieving the goal of a single unambiguous profile for manufacturing worldwide. It is being performed by close liaison with standards bodies and by publicizing the Implementation Guides. Releases of the Implementation Guide are available from the European Commission.

PILOT DEMONSTRATION FACILITIES

The CNMA consortium has delivered a number of pilot demonstration facilities:

* a cell demonstrated at the Hannover Fair, West Germany.

* three production facilities at:

   a) Bavarian Motor Works (BWM) in Regensburg, West Germany
   b) British Aerospace in England, and
   c) Aeritalia in Turin, Italy.

* two experimental facilities at:

   the University of Stuttgart - ISW
   Renault, near Paris, France.

* two production facilities at:

   Aerospatiale, near Paris, France and
   Magneti Marelli, San Salvo, Italy.

The Hannover demonstration provided the first public display of the interworking of hardware and software in compliance with a CNMA Implementation Guide. The objective of this demonstration was to display true interworking, using minicomputers and controllers, connected through CNMA industrial local area networks. The cell was designed to be typical of manufacturing cells; it comprised a machining center, a robot, a transporter and manual stations. With this system, machining of real parts was possible.

For eight days the cell was demonstrated and the aims and achievements of the CNMA project were explained. This was the first demonstration of multivendor interworking using MMS.

Validating new communications software is done by using it in real production activities, where it must provide all essential functions with very high reliability. The CNMA project has validated its communications software in three such environments; at BMW in West Germany, British Aerospace, and Aeritalia in Italy.

BMW's position in the highly competitive automotive industry, means that the company must stay at the leading edge of technological developments to produce ever greater efficiency. They were one of the first companies to recognize the benefits of adopting standards for communications over these networks. This led to their involvement in the CNMA project.

This CNMA production facility is at BMW's new factory in Regensburg, West Germany, where 3-series limousine and convertible cars are manufactured.

On the production line, car assembly tasks are controlled by well over one hundred PLC's which pass any disturbance or error reports to a Siemens Sicomp minicomputer. This information is then relayed to a Nixdorf Targon Minicomputer over a Baseband LAN and a factory backbone Broadband LAN. Information regarding major disturbances is relayed over these networks immediately, using CNMA's MMS message transfer protocols, while less urgent information is transmitted at the end of the production shift, using MMS file transfer protocols.

This comprehensive maintenance scheme is the world's first production application of MMS. It indicates a major industrial company's confidence in the software developed within the CNMA project, particularly since the facility involves just-in-time production. The result is that since early 1988, this software has been used in the manufacture of cars which are a status symbol throughout the world.

CNMA's second production facility is at the British Aerospace factory in Samlesbury, England. This facility machines 'D' shaped components for the leading edge of the A320 Airbus wing. This factory participated in the Enterprise Networking Event, and was the event's only true production facility. CNMA's participation has allowed the project's vast experience in communication software development to be applied to the evolution of the MAP specifications. The experience that CNMA partners gained through the Hanover Fair enabled them to achieve testing for Enterprise ahead of schedule.

The precision boring machine, the centerpiece of the facility, is served by an automated transporter, and is supported by a number of manual process areas. The complexity of the control system allows the facility to be operated as a manned flexible manufacturing system for just-in-time production.

As with the Hannover Fair demonstrator, CNMA has used more computers and controllers than necessary, simulating a large scale CIM installation.

The vendor's devices jointly control the precision boring facility, using CNMA software to communicate with each other.

An advanced feature of the CNMA communications software used in this facility is the Network Management Service. It allows the user to configure and monitor the performance of each layer of the communications software in each device. CNMA was further involved in Enterprise by providing a major part of the required MAP test system, through the third phase of the project, named CNMA Conformance Testing, or CCT. These test tools, which are necessary to ensure interoperability of devices, enabled the Enterprise Networking Event to proceed. This conformance tool set is now marketed worldwide by SPAG-CCT, a Belgian company established specifically for this purpose.

CNMA's involvement in this event, demonstrates that the same software standards are being adopted on both sides of the Atlantic. This is a major step towards enabling manufacturers to integrate devices from different suppliers, European or American, to create a computer-integrated manufacturing environment.

The ability to transfer manufacturing data between companies over a computer network will become an increasingly important feature in advanced industries. For example, companies frequently wish to send updated manufacturing data to a subcontractor. In November 1988 at the Aeritalia factory in Turin, CNMA commissioned its third production facility, integrating minicomputers from Olivetti and Bull. This facility is used in the manufacture of aircraft wire harnesses. As an enhancement to this pilot facility, an X.25 network connection was established between the British Aerospace pilot and Aeritalia. This Wide Area Network link allowed the transmission of wire harness manufacturing data from British Aerospace to Aeritalia, demonstrating communication with a sub-contractor.

More recently, a highly sophisticated experimental facility has been commissioned at the University of Stuttgart. Within the facility, ISW has used equipment from Bull, GEC, Nixdorf, Olivetti, Robotiker and Siemens, along with applications software from Alcatel-TITN to build an impressive demonstrations. The ISW facility consists of two independent cells, one designed to manufacture bodies for hydraulic valves. This cell features a turning center, a boring and milling center, a linear portal robot, and a pallet store. This cell is fully automated. The second cell consists of a manually loaded 5-axis milling machine, connected by a local area network to a component design facility consisting of CAD and NC programming systems. Once part programs have been created, they can be downloaded to the NC controller and started by the cell controller.

The next facility to be commissioned will be at a Renault factory in Boulogne Billancourt, France. This pilot provides a test and demonstration facility for Network Administration systems. In large networks, automated facilities are required to handle faults, tune the network for optimum performance, and manage reconfiguration of the network. Suitable software for this is provided within the network management function. The facility consists of a conference room connected to a test laboratory. Three demonstrations are planned:

    * A simulation of the Aerospatiale pilot.

    * Control of video equipment and the communication of video images.

    * Management of faults achieved by use of a fault injector for
      simulation of typical "real world" faults.

In combination, the facility will demonstrate the ability of the network administration system to tune and reconfigure the network and diagnose and manage fault conditions.

At this stage in the project, a turning point is reached. The emphasis moves from looking for technical confidence in open systems solutions to assessing the business benefits that can be achieved using stable OSI implementations.

The first facilities which can be assessed in this way are:

    * Aerospatiale, near Paris, France.
    * Magneti Marelli, San Salvo, Italy.

The Aerospatiale facility uses CNMA implementations for the integration of a machine shop. This shop is used for the production of prototype missile components. There are nine machine tools in the facility and the objectives of the facility are improvements in quality, flexibility and delivery times. The CNMA communications software is used to integrate four applications. These applications control all aspects of cell control, such as maintenance management, shop management, scheduling and transport control.

At Magneti Marelli, the CNMA software is installed in the final section of an alternator production line. Three primary functions are performed within the facility. These include shop floor monitoring of machine productivity and performance, a tracking system to trace all work in progress through the shop and a diagnostic system to collect detailed information on machine status.

These facilities use the very latest implementations of International Standards, the core services of which are now stable.

CNMA COMMUNICATIONS PROFILE

Communication in CNMA is based on the Open Systems Interconnection basic reference model (OSI/RM), which is defined by ISO in IS7498. It defines a framework for communications, involving the services to be provided to application processes, the breakdown of the communications software into 7 layers and the split of services between the layers. The model is known as the "OSI seven layer model."

Each service requires protocols - specified interactions - between the two systems. Definitions of the services and protocols in each layer are provided in individual standards documents. However, a given service can be provided by a number of different protocol combinations in the lower layers. An additional document is required to identify the exact protocols used at each layer. Such a selection of protocols is known as a "profile."

The CNMA Implementation Guides Defines the communications profile chosen for use in the project. They represent a considerable amount of work by the participating companies and their publication remains a primary function of the consortium.

The main purpose of CNMA is to focus its research upon application layer (layers 6 and 7) issues. This was also a major topic in MAP developments and has been addressed in MAP V3.0. CNMA has contributed to the MAP evolution through its research and implementation work in this area, so aiding the definition of a single international profile of communications for use in the manufacturing environment. CNMA is making strenuous efforts to maintain compatibility with MAP, and also with the standards supported in Europe via such groups as SPAG, the vendor Standards Promotion and Application Group, and through the standards bodies such as CEN/CENELEC.

The structure of the Implementation Guide is as follows:

Volume 1: Transport Profile

The transport profile addresses the issues of network connection, network addressing and reliable data transfer. This volume encompasses the lower four layers of the ISO model.

Volume 2: Application Profile

The application profile contains mechanisms for the control and passing of messages between applications. Specific protocols and services such as MMS, FTAM and Directory Services are addressed.

Volume 3: Application Interfaces

By standardizing the interface to the communications software, the portability of user applications is greatly improved. The FTAH and MMS infaces are described here.

Addendum 1: Network Management Application

During the current phase of work, CNMA has expanded much effort in the area of network administration applications.

Addendum 2: Technical Reports

This addendum is an analysis of certain CNMA related topics, such as Fieldbus, Remote Database Access and Higher Performance Architecture.

CNMA Implementation Guides devote a chapter to each layer or service.

For layers 1 and 3, CNMA initially uses Local Area Networks (LAN's) and acknowledges the benefit to users of providing a choice of LAN type. This allows a user to choose a type based on: cost, performance, installed base, maintainability, etc. Three options are specified, and all have been successfully utilized within the project.

MAP initially opted only for the broadband technology and extended this to cover carrierband with token bus access. Studies in Europe showed that 98% of European LAN installations have opted for Baseband technology. CNMA's choice of Baseband LAN is the same as that supported by TOP.

In CNMA, layers 3 to 5 are designed to conform with MAP. These layers are more stable than layers 6 and 7, and are therefore defined as "background" for the project. This enables work to be concentrated on the other upper layers.

For layer 6 (the presentation layer), which was absent in MAP V2.1, CNMA initially specified the use of a kernal subset. Layer 7 protocols were covered in a number of chapters. The first defined the Association Control Service Elements known as ACSE, which used the latest ISO draft international standard protocol to provide association control.

A chapter for layer 7 protocols defined the Manufacturing Message Specification - MMS, which is a protocol for passing messages between computers, numerical controllers and programmable logic controllers. This was the greatest area of activity in the project.

It is not necessary or desirable for vendors to implement the complete MMS service, so subsets were selected to provide the functionality required for the demonstrators. Typical services were selected as follows:

   * Up and down-load programs to PLCs and NCs
   * Report change of status from NCs and PLCs
   * Start and stop program execution in PLCs and NCs
   * Request status information from NCs and PLCs

In all of these services, the NC or PLC interacts with a computer. For the definitive list of supported functions, the Implementation Guide should be consulted directly.

Another chapter in the layer 7 group presents FTAM. Currently file transfer and management of file structures, is supported.

A further chapter for layer 7 covers network management. This is a protocol which permits a remote device (manager) to access communications related attributes (counters/timers) in other devices (agents). Using this protocol, a manager can monitor and influence network performance, configuration, etc.

Finally, a chapter in this series covers directory services. The protocol and resultant services allow devices to interrogate a remote directory database. This can be used, for example, to establish the address of an application it may wish to transfer a file to.

The CNMA profile is not yet firm, but evolves as the upper layer standards mature. The latest issue of the Implementation Guide produced by ESPRIT Project 2617 is version 4.1 and incorporates the latest ISO specifications including:

* MMS and MMSI,
* FTAM and FTAMI (File Transfer, Access and Management),
* Directory Services, and
* Network Management.

In the current work, network administration is seen as a vitally important service for large computer networks. Consequently, network administration is an interesting market area for the vendors to exploit. The tasks that the CNMA network administration system is designed to perform include:

* Configuration management
* Performance management
* Fault management

By use of knowledge-based techniques, the network administration system can carry out these tasks in an autonomous manner.

The business objectives are to reduce downtime, optimize network performance, and simplify the maintenance and support function. These functions are described in Addendum 1 and are Network Management applications rather than protocols.

CONFORMANCE TESTING

If industry is to reap the benefits of open systems, it is essential that conformance tests are established worldwide which provide consistent and uniform results. This is a necessary step towards ensuring interworking of products from different suppliers.

In the first phase of the project the Fraunhofer Institute, an independent testing organization, obtained and developed the conformance test tools and procedures to verify the CNMA implementations. For testing MMS and ACSE, they developed a test system on a minicomputer system programmed in 'C' and running under UNIX System V.2. The tools developed in this environment are highly portable, permitting future delivery to vendor or user sites, and to other test institutions and developers. This was the first such tool developed for the MMS application protocol which is the main feature of MAP 3.0.

For testing the lower layers of CNMA implementations for the Hannover Fair demonstration in April 1987, Fraunhofer obtained the MAP 2.1 test-bed from ITI in Michigan. These tools were the only ones recognized by the U.S. MAP User Group for MAP 2.1.

During the prestaging for the Hannover Fair, the test tools were used to great effect, testing the vendor's new implementations. The outstanding success of the demonstration is evidence of the quality of the test tools developed so far.

The urgent requirement for test tools for MAP and TOP at ENE '88 provided a rare opportunity for Europe to establish and own a significant test site by bringing together the necessary resources under the ESPRIT program. Besides the existing CNMA partners, SPAG Services, the Fraunhofer Institute, The Networking Center and ACERLI became involved. Agreements with the Corporation for Open Systems (COS) in the USA have been made to ensure the widest possible acceptance of these conformance tests.

The third phase of the project, CNMA Conformance Testing developed conformance tests for MMS Network Management, Directory Services, Bridges, Routers, End System/Intermediate System Protocol and Logical Link Control Class 3. These tests were supplied to Enterprise staging areas. SPAG established a Belgian company, SPAG-CCT SA, that successfully completed product development, and the tools are now marketed worldwide. Further work has been sponsored under ESPRIT II in the form of EP 2292, Testing Technology for CNMA and is known as TT-CNMA. TT-CNMA is coordinated by SPAG, and is furthering the development of conformance test tools. TT-CNMA is also leading the development of interoperability test tools, used to analyze the interoperability characteristics of conformance tested devices. These "IOP" tools can also perform load testing and log protocol interactions. All these aspects enable the integration team to establish confidence in the communications subsystem before installing applications software.

CONCLUSION

In conclusion, the major achievements of the project can be reviewed as follows:

* Four Implementation Guides have been produced defining a communications profile to allow MMS automation protocols, FTAM, Network Management and Directory Services protocols to be exchanged between multivendor systems of minicomputers and programmable devices.

* Standardized user interfaces have been developed for MMS and FTAM to increase portability of application software.

* Network management applications software has been developed to allow superlative control of large networks to be achieved.

* Multivendor control using CNMA communications software has been successfully demonstrated at the 1987 Hannover Fair, the Enterprise Networking Event, and in real production environments throughout Europe. A number of world "firsts" have also been achieved during these activities.

* CNMA can liaison with standards bodies and is having an impact on the final, single communications profile, thanks to its experience in implementation and validation of the CNMA profile.

* The project has brought many European vendors closer to the final standard.

* A comprehensive set of MAP 3.0 test tools are now marketed world-wide by SPAG-CCT, a commercial venture launched specifically for the marketing of the conformance test tools.

* CNMA vendors market OSI based products. These MAP compatible products include minicomputer interfaces, programmable logic controllers, gateways and bridges.

The major impact the project has had on emerging standards is already being built upon with new work extending into the 1990s. This work will be funded under an ESPRIT II contract EP 5104. This will ensure that ESPRIT CNMA continues to make a major contribution to the lower cost CIM facilities that manufacturing industry so urgently needs.

REFERENCES

(1) CNMA Implementation Guide 4.0.
(2) Industrial Local Area Networks: User's Needs.
(3) CNMA Strategy Document.

The Role of CAD/CAM in CIM

by Gary K. Conkol
NIST Great Lakes MTC

INTRODUCTION

People who use similar equipment have organized into formal user groups, focusing attention on issues of fulfilling the potential of advanced technology. Formal representatives of these groups were invited along with a representative of academia and consultants during AUTOFACT '89 to discuss issues central to the use of CAD/CAM in CIM enterprises. This group formally represented over 150,000 end-users of CAD/CAM systems. Seven major topic areas were established from questionnaires and feedback from CASA/SME's technical council, and information received from NIST (The National Institute of Standards and Technology), formally National Bureau of Standards, and user groups themselves.

CAD/CAM is the granddaddy of CIM. Many could argue it is the basis of all modern day CIM efforts. These are two of the original applications of computers to the process of manufacturing. These consisted initially of simple line drawings and fundamental numerical control of machines.

As CAD/CAM systems took shape in the late 1960s and early 1970s, fierce competition developed between manufacturers of "turnkey" systems and systems that contained everything bundled together. Loyalties to particular vendors often dictated the system to buy, and companies became associated with just one vendor. As time went on, differences within companies, and lack of delivery on promises, caused many companies to buy different systems on subsequent purchases. As computers became cheaper, faster, and available in a variety of forms, software proliferated at an exponential rate.

Today, it is hard to find even a small manufacturer who is only concerned with one type of system. If one plant uses primarily one type of system, another plant may contain a different system. Even a company with one type of system is concerned with transfer of their data, either to or, from a dissimilar system. The rate at which computer technology is advancing has caused industry wide argument over which type of system is best.

Yet, in the nearly 30-years CAD/CAM systems have been around, while the fundamental processes of manufacturing may have changed form, the underlying reasons why those actions are done have not. Companies who have changed systems over the years, and those who find themselves bewildered when it comes to choosing a system, are finding stability in re-examining their needs in a generic sense. They are finding that there are common issues, regardless of which system they have, or are considering.

The seven topics discussed by this group included:

1. The state of the technology in CAD/CAM systems
2. A professional organization's role in realizing the benefits of CIM
3. Common ground issues
4. Organizational issues
5. Certification pros and cons
6. Product definition data specifications, PDES, IGES
7. Networking

In addition, an eighth subject was introduced for general consideration. As it turned out, the discussions generated a solid conviction to continue the group meetings. The eighth topic then, was to decide in what format to continue these meetings in.

What follows is an examination of each topic. In each case, the subject will be detailed, and the background, consequences, and analysis presented. In some cases, no consensus was reached, but the discussion itself will cause the reader to form an opinion. The unabridged version of these comments appear in the CASA/SME Bluebook of the same name.

TOPIC #1 -- THE STATE OF TECHNOLOGY IN CAD/CAM SYSTEMS

"What one thing would you like to see to more fully realize the benefits of CAD/CAM?" This was the question asked and sent to the participants in advance of the meeting. The topic was designed to bring out the real issues, and what we must concentrate on to take advantage of the advances we know will come.

Each of the participant's background encompassed vast experience. The advantage of years of experience, in this case, was that technology had changed in at least three major phases during this time frame. From mainframes to minis, from storage tubes to high-resolution raster, and from minis to micros, these people have seen it all. The advantage of this was that they would all be knowledgeable about the subject and could discuss reasons for the changes. What caused the change? Was it push or pull? Is change indicated? Is change in any particular area beneficial, or is it just change for the sake of change? The answer was not as important as whether there was a general direction noted among the group. The bottom line was that a solid direction was noted.

THE RESULTS

The discussion quickly revealed three major trends affecting the result of computer-aided manufacturing, including all activity related to a manufacturing enterprise.

A PRODUCTIVITY PLATEAU

We are experiencing a productivity plateau in the top-end systems. After processor speed reaches 10-50 MIPS and disk storage passes 100 MB at access times of 20-30 microseconds, there is a definite plateau in the productivity using traditional techniques and algorithms. Note that I mentioned "traditional" techniques.

CONTINUING PROLIFERATION OF SYSTEMS

There was unanimous comment that the world we work in is heterogeneous. No one vendor, or set of software handles the entire process. This means there is no one item that could cause a breakthrough in CAD/CAM. There are always requests for cheaper, faster software, able to do more. The number of things a system is called on to do, however, is usually much less than what it is capable of doing. The first class of responses concerned ways of obtaining more of the hidden potential of the hardware and software.

USERS ARE ADVANCING TO DRIVE DEVELOPMENT IN CRITICAL AREAS

Users today are more sophisticated than in the past, and will play a significant role in software development of successful systems. Systems that continue to develop in a vacuum will quickly fall behind software that is "user driven." As such, requested developments were prioritized in relation to the need in industry. Better integration topped the list, with standards seen as the facilitating agent required. When integration was mentioned, the group collectively meant large-scale integration on an enterprise level. It is not enough to integrate analysis with drafting and N/C. Today, integration is needed between marketing, product development, shop floor operations, MRP, order processing, and finance. There are current efforts to involve the end-users in the standards development process, and some CAD/CAM companies actively involve their user groups in the development cycle.

Communications networking is seen as one of the enabling technologies that must grow to fill the need as the enterprise processes synchronize. Some of the other technologies mentioned included faster, more complete methods for translating raster images to vector representations in databases or intelligent scanning. Processor speed and faster display processors were also favorite requests.

ANALYSIS

Consider the impact of the productivity plateau. Using traditional techniques such as the wire frame model, there will be a finite level of technology necessary to optimize the procedure within the human condition. For example, in generating a model of a coffee cup, when it comes to adding the arcs, handle, generating the drawing of the resolution, and processing speed, and input, even the fastest typist or mouser can't task a 50 MIP machine. Resolution is important, but what value is a monitor of greater than 4,000 lines, or a dot less than 0.0005 wide? Is it a limiting factor to be able to display a million colors? It is clear that, given a design technique, there is a finite level of computer technology beyond which any improvement is wasted.

There is also a minimum level of technology that makes a particular technique practical. Take finite element processing on the PC as an example. Without the faster processors and accelerators available today, use of a solver on a PC is impractical. Development of the new processors has been fast, but the development of alternate techniques is more fundamental. Techniques used to form finite element stress analysis were actually developed to form the deflection equations for beams back in the 1930s. New methods are much slower to develop than hardware advances, but the important implication is that continued hardware development will cause techniques thought to be impractical to become widely used.

The major impact of the first conclusion, then is two-fold: first, the current confusion and hype over which is the best processor and display system will diminish. More attention will be focused on the technique used to arrive at a desired end. This is the second and more profound implication. We are entering an age where development will be applied to modernizing design, analysis, and production techniques. This makes the third trend mentioned extremely important. Established and traditional techniques will have to be critically and continually reviewed.

The second trend of proliferating systems seems to be in conflict with the idea that processor hype and confusion will diminish. Actually, it makes perfect sense and implies that standards will be more important than ever. As hardware designs become better, they will quickly establish themselves, be copied, and manufactured by an increasing number of vendors. Systems will proliferate with the effect of decreasing cost per capability values. Interconnection and interoperability standards will be paramount to the successful use of these systems.

The third trend of increasing user involvement will be the long-term effort. This is the area where new methods of stress analysis, solids modeling, and machining will come from. These new techniques will offer orders of magnitude improvement over the current system, and will change the face of many industries. The greatest user involvement is found in user groups.

TOPIC #2 -- A PROFESSIONAL ORGANIZATION'S ROLE
IN REALIZING THE BENEFITS OF CIM

Organizations such as CASA/SME are continually searching out the best ways to advance the position of their members. Many of the participants belonged to several organizations. This topic was arranged to find the areas in need of help from the user's point of view, and how it should be done to help the people who must solve the problems every day. The one thing all the participants had in common, with the exception of the consultant, was that all were members of a user group. They had all experienced operating within an organization designed to facilitate the member's success.

THE RESULTS
USER GROUP GOALS AND OBJECTIVES

Obviously, user groups were unanimously supported. Each member brought out their organization's goals, and how they pursued them. Basically, the overall goal was communication. By prompting communication between users, and vendors, they were able to drive the offerings closer to their needs. Vendors were seen as generally receptive to this, and in most cases took direct input from their groups into the development cycle. Without exception, the national groups, augmented by local and intracompany groups, cited success after success in helping users get more out of their systems. It was also noted that the most active companies in the user group organization were people who got the highest productivity out of their system. This could be cause or effect, but the fact that user group activity is the most cost effective investment possible was clear.

CASA/SME GOALS AND OBJECTIVES

The discussion was shifted to an nonaligned organization such as CASA/SME. Half the group had not been involved with the organization, but knew of its existence. All had seen the CASA/SME CIM Enterprise Wheel and knew of its concept of integration. CASA/SME's value was seen as growing from a common neutral ground in which to pursue developments in the interest of all the groups. CASA/SME has the potential here to collect the important issues and topics and to disseminate them to all vendors, as well as focus national attention of the real issues affecting all users and companies in the pursuit of CIM. In activities related to common goals, the user groups could be expected to share an active role with CASA/SME.

FORUMS OF THIS NATURE

The Round Table discussion method conducted by CASA/SME was judged as an effective tool for furthering the state-of-the-art and market. Interestingly each group stated many of the same needs and frustrations. As more and more of the same comments emerged, the energy and conviction grew. Not only did the participants want to continue on with subsequent meetings, but expansion was requested. The formats discussed included the possibility of establishing a mini-conference. Everyone agreed, however, that they had enough conferences to attend and that travel money was always in short supply. A one-day expansion to an existing conference such as Autofact was offered as a suggestion. It was clear that enough good had already come out by getting the groups together on neutral ground, that regularly scheduled meetings were mandated.

ANALYSIS

It was interesting that the relationship between users and organizations such as CASA/SME was minimal. Only half the people in the room had been to a CASA/SME meeting, and the rest were not present on a regular basis. The reason involved travel, expense costs, and time away from work. Closer coordination was requested, and the forums became the preferred method. CASA/SME and organizations like it were considered neutral environments within which to formulate plans to further the cause of all users.

Further analysis of the recommendation is not necessary. The format of the meeting has started, and subsequent meetings are planned. There was much discussion on whether to include the vendors in these meetings. The general feeling was that the vendors should be included for part of the discussions, but not all. The major impact here is that it now falls on CASA/SME to collect the energy of these groups, formulate, summarize, and take action on the recommendations. This is clearly within the organization's charter and is welcomed by the users.

TOPIC #3 -- COMMON GROUND ISSUES

The topic "common ground issues" opened up the floor for each contingent to stress their own priorities. Each participant was asked to consider the issues they thought were important to the industry. Expectations ranged from processor speed to capital funding availability. What we found should not have been a surprise.

THE RESULTS

When asked what common issues all users of CAD/CAM faced, we were all mildly surprised to find that everyone, without exception, had the same first choice. Training was the issue with the highest priority. In fact, only one other issue was mentioned. This would not be surprising, except that the group was asked to come up with the list of issues independently. In summary, the responses speak for themselves, but there were four basic types of training needed:

1. Management training, which refers to enlightening upper management on the capabilities of modern CAD/CAM systems and how they are (or should be) used in the enterprise.

2. End-users should be continually trained in the technology available in CAD/CAM systems. Current examples would include scanning, photorealistic rendering, file transfer standards, and modeling techniques.

3. Specific vendor training was a requirement to get the most out of any system, but was only part of the answer.

4. Training is mostly required in applying the vendor's capabilities and the technology to the job at hand within the enterprise. Many are familiar with certain vendor's claims that the system can be learned in one week or less. That claim may be true when measured by the user's knowledge of the commands, but is off by months when measured by the ability of the user to apply the system effectively within the enterprise.

ANALYSIS
PEOPLE ISSUES AND HUMAN NATURE

The broadest issue in realizing more of the potential of these type of systems was a human issue. Better education in the technology was cited as a prerequisite, rather than specific product knowledge. People must understand what it is they are trying to do, rather than be intoxicated by the features of a system. The way to do this is to get universities to pursue aggressively the fundamentals associated with the use of CAD/CAM. Technology transfer organizations such as the NIST centers will also be key to passing on the techniques independent of the product. Better focus is needed in the lower educational grades on what engineering and manufacturing is all about. Many states have programs, such as the visiting scholar program in Ohio, where experts from industry travel out to grade schools and high schools to give students a flavor of industry. More focused education is needed.

MANAGEMENT AND PROCEDURAL ISSUES

Another recurring theme was related to management and procedural issues. The old way of doing things must be critically re-examined. There is an old science fiction movie entitled "Forbidden Planet," where a race of aliens had perfected the ultimate machine to manufacture anything they thought of. Man stumbled onto the plant centuries after the race had become extinct, only to find they had succumbed to what was found to be "monsters of the id." The "id" refers to the subconscious. What happened was that the aliens had subconscious desires to harm each other. The new machine merely complied and annihilated the race overnight. What we find here on planet earth is that we have our own "monsters of the manufacturing id." We find that modern systems are often constrained by the old way of doing things to the point that we lose the benefits we sought to gain. We must continually ask ourselves if we are using a procedure because we always have, or because it is essential?

TOPIC #4 - ORGANIZATIONAL ISSUES

This topic asked each user group to describe their organizational structure and discuss how it worked. Remember, the people commenting are either leaders or other representatives of groups which have been in existence for over ten years. This test of time in the turbulent market of CAD/CAM is a definite validation of an organization committed to helping its members succeed.

THE RESULTS

The direction among user groups was to become more involved in the development and integration of their systems. This includes direct input into the marketing departments, as well as the engineering departments of the vendors. The groups also universally supported local extensions to the national organizations.

In terms of organization, most user groups were organized along the lines of applications, with special interest groups for each major demarcation. A few special interest groups were organized for special or unique products and hardware offerings peculiar to the individual vendor. The structure was almost identical among the groups, with an executive board consisting of various chair persons, a president, vice president and treasurer, as well as some sort of liaison with the vendor themselves. All the organizations relied on volunteers entirely, with some paid services, such as legal and financial.

ANALYSIS

The structure of user groups is very nearly a standardized model. We also found that the human network supported through the user groups is very extensive, commonly reaching thousands of people. In one user group, the chairperson had set up a "fanout" calling procedure, where each member of the steering committee would receive a call and then call two others. Through local user groups, it was possible to reach the entire nation in less than a half hour. A question could be asked and a national poll could be taken in less than two hours, while reaching the majority of users.

User groups are here to stay, even with the decreased dependence on the turnkey vendors. The users rely on these groups to find the answers to questions no one else can answer. The cost to a company in supporting user group activity is well justified.

TOPIC #5 -- CERTIFICATION

Certification is a hotly contested topic, due primarily to three main reasons which came out in discussion. The first, and perhaps the most evident, is that a certification rating implies a premium in the expected compensation. Whether or not a company gives a premium for CAD knowledge is minor compared to the expectations and polarization this aspect generates. Most people would rather not recognize these programs rather than institute a system that may be complicated to administer and unfair to those who do not test well.

The second reason behind the difficulty in dealing with certification is the validity of a test. The cost of generating and maintaining a quality test easily exceeds six figures and is often limited to a specific industry.

The third reason is that computer systems and features change so quickly that what is important to know becomes an ever difficult target to hit. The number of vendors in the market, the types of computers available, and the options, all make the generation of a test extremely difficult.

THE RESULTS

It was predictable that there was no outright agreement on this topic. The overall conclusion to discussion on certification was not to pursue a formal CAD operator class, but to review the possibility of certifying CAD/CAM managers. These would be the people charged with making effective use of the systems within their companies.

The need to quantify an operator's proficiency would be best satisfied by organizing a series of "top gun" contests at the user group level. Tests could be quickly generated and revised every year, as well as fine-tuned to the highest productivity applications. The tests would be optional and the dissemination of the results up to the individual taking the test. This would give the user a mechanism to rate himself and quantifiably brag about it to his superiors.

ANALYSIS

With all the negative aspects of certification, why is the topic still discussed? Why isn't certification dead? The answer is simply need. CAD designers and operators need a way to rate themselves. Management needs a way to determine the qualifications of new employees and how well users are doing within their company. Educational institutions need a way to develop curriculums and tests. The fact is that even though it is difficult, there is still a need to identify those skills which distinguish a person who will be successful using a CAD/CAM system. Certification may not be the answer, but some form of testing is definitely indicated.

TOPIC #6 -- PRODUCT DEFINITION DATA SPECIFICATIONS

The sixth topic involved standards and data transfer. Specifically, each person was asked what they thought of the international efforts underway in IGES and PDES, and if the work should continue. While the casual observer might think the answer is yes, the comments from those involved in trying to use the standard in the workplace raise some questions and call for a recommitment to practical solutions.

THE RESULTS

The overall result of this polling indicated a serious problem in the area of design data transfer between systems. Example after example of problems were noted. In a time-critical environment, the solution is often to abandon the best method in favor of the most expedient one. Direct translators with plenty of development behind them, or application specific translators were used most. In most cases, the use of direct translators was discouraged, due primarily to the costs, risks, and maintenance overhead. The success stories of using the standards such as IGES relied on "flavoring" techniques, which prepared the model before and after translation. Most users were not involved in the volunteer organizations such as the IGES PDES Organization (IPO) and were unsure how to make their views known.

ANALYSIS

The first thing that should be noted is that there is a growing need for seamless data transfer between systems. The usage is there and growing; unfortunately, so are the problems. The second item that has to be noted is that the standards are seen as the hope to solving this problem. While users may criticize the existing standards, they also realize that if this is ever to work, it will require strong comprehensive standards.
Two additions which should have significant impact in this area are the generation of case study documents from the IPO organization, and the formation of an IGES user group. There are numerous examples of how users either solved or circumvented a problem in data transfer between systems. Up until now, how they did it was often not recorded, and if it was, it was not in a format that others could use or discover. The IPO organization is seeking to remedy this by documenting, filing, and disseminating case studies of IGES file use. A library has been established, and NIST will be the repository.

The IGES user group will disseminate information regarding the use of the standard. Until recently, this information had to be gathered from the standards development activity within IPO. This slowed down the process of developing the standard and often bewildered new attendees. This group is currently forming through the National Computer Graphics Association in cooperation with others.

The impact of this topic extends to all systems. It has already been noted that the world of CAD/CAM is heterogeneous, and growing more so. Standards will become more critical in the daily use of these systems. Migration to the PDES and STEP standards will fulfill capabilities that are required by modern systems, but the degree of complexity increases accordingly. While development of these standards is proceeding, the amount of work remaining and the urgent need are not on the same time line.

TOPIC #7 -- NETWORKING

Networking was added to the list of topics due to its pervasiveness in installations everywhere and its growing popularity. We have already seen that most installations have multivendor distributed architectures, or are going that way. Within the past few years, the inclusion of the PC into the mainstream of the enterprise has been facilitated by networks. Today, networking is effectively tying together computers, but what about the factory floor? The imminent storm over broadband MAP and OSI compliant protocols is about to shake the foundations' network architecture. While it is one more transition that everyone fears, it will also provide the connectivity required to truly integrate the enterprise in real-time.

THE RESULTS

Most users employed networks to save time and, as long as they worked, didn't pay much attention to them. All agreed, however, that networking was a major issue. Many cited the different networks they had installed and the different functions they were used for. Standards once again became the center of discussion. At least networking standards could be monotomically optimized for speed, unlike standards for computer graphics which vary with application. It was recognized that high-speed networks enable computer resource sharing, which will require new operating system capabilities. This change was not fully embraced, due to its impact on hardware and software systems already installed. All agreed it was necessary.

ANALYSIS

As discussions progressed, it became clear that the major issue was integration within an enterprise. Networking solved many problems within a department, but then brought attention to incompatibilities between departments. Once data transfer becomes easy between different functions, it is quickly found that the data content is incompatible or discrepant. This may not be good news, but solving these incompatibilities is a prerequisite to integrating the enterprise, which has been long known to yield the gains necessary to maintain competitiveness in the future.

EPILOGUE

By the end of discussions on the seven prepared topics, the group was literally amazed on how well this meeting of diverse interests went. In fact, the interests were not diverse at all but common, and even succinct. As each representative was asking for one last comment, there was a spontaneous reaction resulting in a call to action. The suggestions and recommendations basically broke down as follows:

* More interaction was requested. Additional meetings of this type or expanded meetings covering one or two days were indicated.

* There was much discussion about whether or not to include the vendors. The overall feeling was that there is a place in the discussions where we want the vendors involved, but not during the entire meeting.

Actions since the meeting have resulted in some groups joining force and, as results appear, efforts will accelerate. This was seen as a step along the path of joint partnerships and open communication, necessary to advance the state-of-the art and realize the elusive benefits of CIM.

Attribute-Based Group Technology System

by P.H. Cohen, S. Joshi, S. Hsieh, H.S. Lu
The Pennsylvania State University

R. Goddard, R. Stimpson
American Sterilizer Company

INTRODUCTION

Since the beginning of the Industrial Revolution, many innovative techniques have evolved to improve manufacturing efficiency. One that has been widely studied is group technology (GT). In 1925, R.E. Flanders presented a manufacturing method to avoid difficulties in manufacturing and production control that today would be classified as group technology (1). Since the popularization of GT by Mitrofanov (2), a great deal of pioneering research has been done along with successful implementation.

FUNDAMENTALS OF GROUP TECHNOLOGY

A definition of Group Technology has been given as follows:

Group Technology is the realization that many problems are similar, and that by grouping similar problems, a single solution can be found to a set of problems thus saving time and effort (3).

The basic foundation of applying GT philosophy to production is the part family, which can be defined as a group of parts with some specific similarity in design and manufacture. Such similarity can be exploited to improve the effectiveness of design and manufacturing.

For product design, products with a similar shape can be grouped into design families. When a new product design is required, an existing product design from the same family can be retrieved and a certain degree of modification can be made to meet the specification of the new product. This saves a significant amount of design time. For product manufacturing, products which require similar manufacturing processes can be categorized into production families. Based on the production families, a variant process planning system can be established. Machines used to process a production family can be grouped to form a machine cell, greatly reducing processing time of products by decreasing part travel time between machines.

CLASSIFICATION AND CODING SYSTEMS

The key to implementing GT successfully is grouping related parts into part families. Coding and classification techniques has been the primary method of accomplishing the task. Coding is a process of establishing symbols which represent the manufacturing attributes of the parts. Classification is a process separating parts into groups based on the manufacturing attributes of the parts. Before a coding scheme can be constructed, a survey of all part features must be completed and attributes to be coded are selected. Then a coding scheme is designed to represent selected attributes, and code values are assigned to the features. Once a coding system has been established, and components have been coded in that system, a classification method is used to form the part families. Usually, a clustering algorithm is employed to produce the desired part families.

A number of commercial and public domain classification and coding (C&C) schemes are available and well-documented. Some use as few as five digits to describe the component, such as the well-known Optiz system of West Germany, while others use 21 digits for the same purpose, such as the KK-3 system of Japan. Since these systems have been developed for general use, they usually do not fully meet the needs of specific users. However, these coding schemes provide a useful base, and a certain degree of modification can be applied to these systems to meet the specific needs of the user.

Recently, the digital computer has contributed a great deal of improvement to many manufacturing technologies, including GT. Obviously, the expansion of computer capabilities and the availability of software has helped the growth of the GT applications. A number of computer-assisted coding programs have been developed within the past several years. A few of them are ICACC (The Interactive Computer-Aided Classification and Coding System) (4), ACAPS (The Automated Coding and Process Planning System) (5, 6), MICLASS (The Metal Institute Classification Program) (7), COFORM (Code for Machining) (8), DCLASS (9), and AUTOCODE (Automated Coding) (10).

Many industries have developed company specific coding systems based on either modifying existing public domain coding system or creating their own. The fixed digits type of coding system still has limitations for the following reasons:

1. They are generally inflexible for adaptation to different part populations, and the capacity of such a system is limited by the number of digits and coding fields.

2. Since C&C systems contain alphanumeric fields to represent discrete ranges, exact attribute values, such as: length, width, and height of a part would not be captured unless they are read from a part drawing.

3. Multiattribute data cannot be coded in a single digit position. For example, an external flat feature of a sheet metal part can be one or more of the following: square corners, radiused corners, chamfered corners, side notches corner notches, simple cutouts, complex cutouts, contour/radii, and/or tapers. It is impractical to develop a fixed digits-type coding system which can represent all possible combinations.

An attribute-based C&C system has all the characteristics of a GT database (DB). Unlike the traditional C&C systems, actual attribute values for each part are stored in database. For example, if a part has a particular length, a specific code value would be used in a traditional C&C system to indicate that the length falls in a specific range. In the attribute-based GT system, however, the actual value would appear in the database. This provides far greater resolution for the formation of part families, design searches, and computer-aided process planning. In addition, unlike traditional C&C systems, unique record retrieval is guaranteed, since part number is defined as a unique field.

GROUP TECHNOLOGY BENEFITS

Recently, benefits derived from group technology application have been reported by the users. Almost every documented case study contains a statement of the significant improvements achieved through the implementation of GT (11-13). Table 1 shows GT benefits from three different studies. The first was performed by the Machinability Data Center (11), the second from Hyer and Wemmerlov (12, 13), and the third is documented by John Deere (11, 14, 15). In addition, Deere & Company has documented over $9 million of cost savings. Estimates place continuing annual savings at over $2 million per year.

Table 1. Group technology benefits.

Savings Category	Average Percentage Reduction		
	Machinability Data Center Study [11]	Hyer & Wemmerlov Study [12,13]	John Deere Study [11,14,15]
part numbers		9.00 (max. 20)	81.48
part design	52.00	22.40 (max. 80)	
time to create a new design		24.00 (max. 75)	
time required to retrieve an existing design		32.50 (max. 50)	
number of design errors		30.00 (max. 50)	
number of designers needed		10.00 (max. 10)	
setup time	69.00		85.00
inventory			60.00
floor space	20.00		27.00
lead time			42.00
machine tool			31.00
work in process inventory	62.00		

GT DATABASE MANAGEMENT SYSTEM (DBMS)

A critical factor in successful automation of a manufacturing enterprise is a well-designed and well-managed database. The AMSCO GT-DBMS provides the basis to do so. The system has a total of 18 modules organized into three groups: raw material, component parts, and commercial items. The data structure which will be discussed subsequently for each module is independently constructed. Each one is a database by itself, and each record has a unique set of attributes or fields capturing the key features of parts. The number of fields in a record among these 18 databases are ranged as low as 10 to more than 100 fields. These 18 modules are grouped as follows:

1. Component parts group: fasteners, plate, round bars, nonround bars, plastic/rubber, castings/forgings, structure shapes and extrusions, and piping/tubing/conduit.

2. Raw material group: contains bill of material (BOM) items for component parts.

3. Commercial items group: electrical, fasteners, bearings, piping/plumbing, bearings, miscellaneous hardware and mechanical, labels/decals, printed documents, packaging material, and other purchased items.

The AMSCO GT-DBMS overview is shown in Figure 1. The kernel GT DB controls all 18 DBs which all are interfaced to the company's MRP DBs. Future expansion may allow the GT DB to also interface to the CAD, Purchasing, and other DBs.

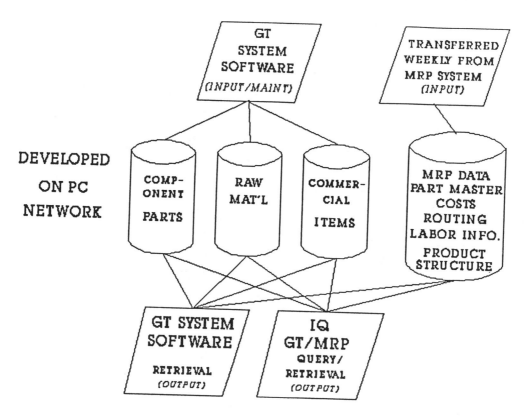

Figure 1. GT system overview.

The GT DBMS developed is an interactive, menu-driven system. To preserve the integrity of the DB, control logic is built-in for most fields. For example, an input value for the "part number" field has to pass the company's standard format before the cursor can move to the next input field. Figure 2 shows a sample input screen for the sheet metal component part module.

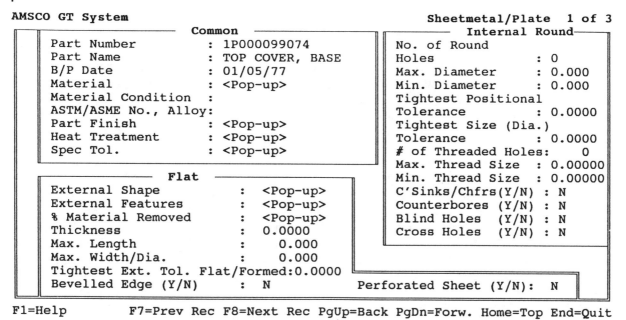

```
AMSCO GT System                               Sheetmetal/Plate  1 of 3
                      ┌──── Common ────┐      ┌──── Internal Round ────┐
  │  Part Number        : 1P000099074  │  │ No. of Round                    │
  │  Part Name          : TOP COVER, BASE │ Holes              : 0          │
  │  B/P Date           : 01/05/77      │  │ Max. Diameter      : 0.000     │
  │  Material           : <Pop-up>      │  │ Min. Diameter      : 0.000     │
  │  Material Condition :               │  │ Tightest Positional            │
  │  ASTM/ASME No., Alloy:              │  │ Tolerance          : 0.0000    │
  │  Part Finish        : <Pop-up>      │  │ Tightest Size (Dia.)           │
  │  Heat Treatment     : <Pop-up>      │  │ Tolerance          : 0.0000    │
  │  Spec Tol.          : <Pop-up>      │  │ # of Threaded Holes:   0       │
  │                    ┌──── Flat ────┐ │  │ Max. Thread Size   : 0.00000   │
  │                                     │  │ Min. Thread Size   : 0.00000   │
  │  External Shape     : <Pop-up>      │  │ C'Sinks/Chfrs(Y/N) : N         │
  │  External Features  : <Pop-up>      │  │ Counterbores (Y/N) : N         │
  │  % Material Removed : <Pop-up>      │  │ Blind Holes  (Y/N) : N         │
  │  Thickness          : 0.0000        │  │ Cross Holes  (Y/N) : N         │
  │  Max. Length        :    0.000      │  │                                │
  │  Max. Width/Dia.    :    0.000      │  └────────────────────────────────┘
  │  Tightest Ext. Tol. Flat/Formed:0.0000 │
  │  Bevelled Edge (Y/N)    :  N    │  Perforated Sheet (Y/N):  N
F1=Help        F7=Prev Rec F8=Next Rec PgUp=Back PgDn=Forw. Home=Top End=Quit
┌──────────────────────── Instruction ─────────────────────────┐
  Move the cursor to each cell and input values or select desired choices.
```

Figure 2. Sample input screen for sheet metal parts.

The GT system developed also provides built-in function keys for all 18 modules. It has the capability to provide on-line help instruction and key words, to automatically generate hard record copies, to interface with other databases from different plants, to automatically write records to the GT DB, to view a family of records efficiently, and to interface with the MRP database.

DATA STRUCTURE AND ORGANIZATION

Database descriptions such as: key, type, length of a record, and size of the DB, are written using the Data Definition Language (DDL), or Data Dictionary Language (DDL). DDL has statements that describe, in somewhat abstract terms, what the physical layout of the database should be.

As a front-end to DDL, a screen management package called ScreenIO (by Norcom) was used. An advantage of using ScreenIO is that it automatically converts user-defined fields into COBOL DDL. The outputs generated by ScreenIO are then linked and compiled with the host language written in the Realia COBOL.

The host language is also interfaced with a commercially available DBMS called Btrieve (by Novell). Btrieve keeps all indices to the data records in the form of a B-tree (balanced tree) structure. Some major advantages of B-tree organization are that desired records can be accessed quickly, can be accessed in sorted order, and can be stored efficiently. When a record is inserted, updated, or deleted, Btrieve adjusts all the indexes for the database to reflect the latest changes.

GT-DBMS OPERATIONS

The object of GT-DBMS is to efficiently and effectively insert, retrieve, modify, and delete part number records. Because the GT DBs are organized using a B-tree structure, the speed for these operations is tremendously fast. For a database size of more than 400,000 records, there is no perceived delay in the retrieval of a desired record based on part number.

Insertion operations insert a record into a GT DB. The records are stored based on the part number. To preserve the integrity of the GT DB, if any newly inserted record that has the same part number as an existing record in the DB, then the record will not be inserted. Modification operations include a deletion (removing of an existing record permanently from a GT DB) followed by an insertion. Similar to insertion and deletion operations, Btrieve adjusts all the key indices accordingly to reflect the last changes. One feature that makes the AMSCO GT-DBMS such a powerful system is the flexibility to retrieve desired records without any knowledge of a query language. Retrieving an existing record based on a unique key, such as part number is the most fundamental operation for any DBMS. If a record is found, all the fields for that record are retrieved and viewed immediately. One of the major advantages of AMSCO GT System is that it requires no user knowledge of special query languages or data manipulation languages (e.g., SQL). For example, using the fasteners database, the retrieval of a record based on a group of representative attributes, the SQL might be written as:

```
SELECT  all
FROM    Fasteners
WHERE   fastener > = 'Rivets' AND
        screw-type > = 'Oval C'Sink Head' AND
        nominal-id > = 1.2 AND
        thread-per-inch > = 2 AND
        overall-nominal-length > = 3.245 AND
        thickness > = .12 AND
        part-number > = ''
```

This SQL command syntax will retrieve a group of records, if any, that satisfies the "where" conditions. The GT DBMS performs the same function. It allows the user to key in desired field values from a "View-Only Sort Data" screen. An example of such a screen for the fasteners module is shown in Figure 3. These field attributes are then internally concatenated into a single string. Appropriate groups of records are then retrieved based on this string.

SYSTEM USAGE AND INTEGRATION WITH OTHER FEATURES

The group technology system will be used by a variety of engineering and manufacturing support groups. The goal of the system is to provide an

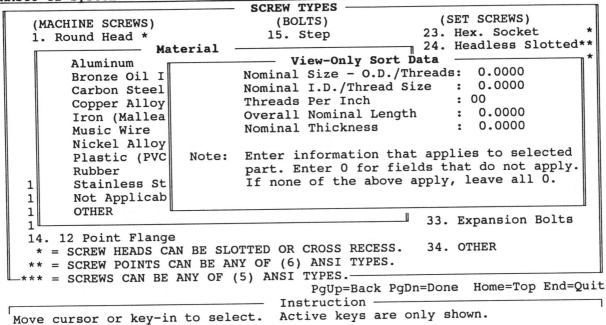

Figure 3. Sample screen to query database about fasteners.

engineering tool that will reduce the total number of designed parts and standardize designs. These goals result in a variety of other benefits and cost savings in each functional area.

The engineers and designers in both our research and development and our current products engineering groups will use the system to retrieve existing designed parts. The intent here is to compress the development cycle for new products through the use of existing designs. This also provides for design standardization and the reduction of engineering costs associated with the development and change process of a new part.

Manufacturing engineers will use the system as a means of standardizing process plans. The GT system will provide an index to our existing process plans or master plans if developed. This will provide consistent process planning which will help to control manufacturing cost, production planning, tooling costs, etc. The system will also be used by manufacturing engineers to provide cost information during the product design process. Manufacturing costs for similar parts to those being designed can be retrieved. This will help to provide early product costing during the product development cycle.

Purchasing will use the system to do analysis work on purchased components. Vendors will be analyzed for consistent pricing. Cost savings can be searched out on high priced parts. In addition, parts can be grouped for family of parts buying, achieving quantity discounts.

Manufacturing management will use the system to control flow of product through the factory. Family of parts batch manufacturing can be better controlled in this manner. The GT system will provide a research tool that can be used to investigate the formation of manufacturing cells. Process and product cells can be developed with GT assisting in the process. The GT system will be integrated closely with existing design and manufacturing systems. For the most part, GT will become the "front end" system used in the design and manufacturing process. Initially, GT will only be linked electronically with MRP databases. It will, however, be integrated with other manufacturing and design systems in a procedural manner.

Engineers and designers will use GT prior to starting a design on CAD. They will seek out existing designs that will meet their need, or one that can be modified slightly for the new design. GT will become the indexing system to all CAD design files. Initially, designers and engineers will interface with two work stations, a PC for GT and a Unix work station for CAD. It is hoped that, in the future, GT queries can be made from the Unix CAD work station. As new parts are designed in CAD, they will be added to GT for future retrieval. In the future, electronic ties may be possible between CAD and GT to automatically pass geometrical information from CAD to GT. Manufacturing support groups, such as manufacturing engineering, purchasing and production planning will use the system in a like manner. GT will be used to supplement or front end their existing systems by finding similar designs, creating part families, etc. The GT system has been developed on a large Novell personal computer network. AMSCO presently has an extensive PC network. The system supports three manufacturing facilities in three different states and numerous regional sales/service offices. Presently the system is loaded on five file servers supporting design and engineering, manufacturing engineering, purchasing, production control, quality control, and our remote manufacturing sites. Quick and easy access to the system can be achieved via any PC attached to these file servers. Remote access across the network backbone or via remote dial-up is also available from any other PC on the network. The base computer is an IBM AT compatible personal computer utilizing the Intel 80286 processor, 1 megabyte of memory, 1 floppy drive, no hard drive, and a monochrome monitor.

APPROACH TO SUCCESSFUL IMPLEMENTATION

The development team was made up of individuals familiar with group technology, as well as manufacturing and engineering from Penn State and AMSCO. The first step was to anticipate most of the potential uses and requirements of such a system throughout AMSCO.

Informational meetings were held with representatives of all departments that would use the GT System. At these meetings, representatives were informed of the progress of the development of the system. They were also asked to make comments and suggestions on the system and their possible uses of it.

At the start of the development program, there were regular meetings between Penn State and AMSCO to define the architecture of the system and to identify part attributes that would be critical for all applications of the GT system. This is a critical phase in the development, since the decisions made at this stage impact the system development and its intended performance. As the system started to take shape, the AMSCO team was able to continue defining the part attributes and provide Penn State with the information needed to develop the programs and system architecture. An electronic mail system was used extensively to communicate between the Penn State and AMSCO teams, as well as to transfer software.

Throughout the program, contact between AMSCO and Penn State has been maintained on almost a daily basis, either with the electronic mail system, or by telephone. This has been a key element in the successful development of the group technology system.

As the programming of each module is completed by the Penn State personnel, the software is electronically mailed to AMSCO. The new software is used to load test parts into the database for the part category under development. Enough parts are loaded to include all possible attribute combinations, testing each possible selection in the software. This approach provides a test of the correctness of the software, and provides an opportunity to evaluate the design of the module. Changes are often made to the modules based on the test.

AMSCO held a seminar for managers, supervisors, and users from all departments to familiarize them with the GT System concept, and inform them that development of such a system was underway. The seminar defined GT, and showed the benefits of a GT system. After several months of development, AMSCO team members traveled to the Montgomery, Alabama plant to present the same basic information to the users at that plant. At this seminar, some of the completed work was demonstrated and feedback solicited on the impression that the potential users had of the system.

When a prototype software module was complete and capable of storing information, a group of manufacturing engineers were asked to "code" parts. After limited training, they were able to load parts into a database using the newly developed software. They were encouraged to evaluate the system for efficiency and effectiveness as they "coded" the parts. It was shown that part attribute information could be loaded quickly and accurately by individuals familiar with part blueprints with an average time of three to four minutes per part. It was also shown that the information could be retrieved effectively to find like parts, or to form part families. When the first working module--Raw Materials--was ready for attribute loading, a team of individuals from various departments was trained to load raw material attribute information into the database. This not only got the information loaded into the database, it also familiarized 16 potential users with the GT system. Afterwards, potential users of the Raw Material module were trained on retrieval of information from the system. The training consisted of an hour and a half to two hours of hands-on classroom instruction, with each trainee at a PC. Each trainee was given a copy of the documentation for the Raw Material module. This documentation had been written as part of AMSCO's in-house development of the system and is also provided on-line within the system as a help facility.

In-house education about GT has been a vital tool in the early success of the program. By involving various aspects of the business (manufacturing, design, purchasing, etc.) in the early phases of development, potential users of the system actively participate in the development of the system that will serve their needs. This approach ensures acceptance and continued use of the system.

SUMMARY

The group technology system developed utilizes a database management approach. This approach has the advantage of being able to store and manipulate a tremendous volume of part data, both quantitative and qualitative, in an efficient manner. The system developed also facilitates simple search and query procedures making the system very user friendly.

Implemented in a network environment, the GT system links directly to MRP, and will soon link other databases at AMSCO. The system is planned for use by manufacturing, engineering, purchasing, production control, management, and others. As a result, it serves the purpose of integrating the various functions of the business.

Close cooperation with the developers and the appropriate consultation with and education of the departments to be effected was the key for the successful implementation and acceptance of the GT system.

ACKNOWLEDGEMENTS

The authors wish to thank the Commonwealth of Pennsylvania through the Ben Franklin Partnership Program and the Industrial Modernization Program for their support in this project.

REFERENCES

1. Gallagher, C.C. and Knight, W.A., "Group Technology," London, Butterworth and Co., 1973, p.5.

2. Mitrofanov, S.P., "Scientific Principles of Group Technology," Translation, Yorkshire, England, National Lending Library for Science and Technology, 1966.

3. Solaja, V.B. and Urosevic, S.M., "The Method of Hypothetical Group Technology Production Lines," CIRP Annals, Vol. 56, No. 4, April 1977, pp. 37-42.

4. Carringer, R., "Interactive Computer-Aided Classification and Coding," Master of Engineering Paper, Industrial and Management Systems Engineering Department, The Pennsylvania State University, 1982.

5. Emerson, C. and Ham, I., "An Automated Coding and Process Planning System Using a Dec PDP-10," Computers and Industrial Engineering, Vol. 6, No. 2, 1982.

6. Emerson, C., Bond, V., and Ham, I., "Automated Coding and Process Selection--ACAPS," Proceedings of the IXth North American Manufacturing Research Conference and 1981 SME Transactions, Society of Manufacturing Engineers, May 1981.

7. Houtzeel, A. and Schilperoot, B., "Group Technology via Computer," American Machinist, Vol. 119, No. 17, 1975.

8. Barash, M., "Group Technology Considerations in Planning Manufacturing Systems," SME CAD/CAM IV Conference, November 1976.

9. Chang, T. and Wysk, R., "An Introduction to Automated Process Planning Systems," Prentice-Hall, Inc., Englewood Cliffs, New Jersey, 1985.

10. Mackowiak, R.M., Cohen, P.H., Wysk, R.A., and Goss, C., "Development of a Group Technology Workstation," The Third International Conference on CAD/CAM, Robotics and Factories of the Future, Southfield, Michigan, August, 1988.

11. Snead, C. S., "Group Technology," Van Nostrand Reinhold, New York, 1989, p.211.

12. Schaffer, G., "Implementing CIM," in "Group Technology at Work," ed. by Nancy Lea Hyer, Michigan, SME, 1984, p.74.

13. Hyer, N.L. and Wemmerlov, U., "Group Technology in the U.S. Manufacturing Industry: A Survey of Current Practices," International Journal Production Research, 1989, Vol. 27, No. 8, pp. 1287-1304.

14. Deere Tech Services, "JD/GTS John Deere Group Technology System Documented Savings," Deere & Company, 1987.

15. John Deere Computer Integrated Manufacturing Services, "JD/GTS John Deere Group Technology System," Deere & Company, 1987.

CIM Implementation: Modeling With an IDEF Perspective

by Captain Michael K. Painter
Armstrong Laboratory
Human Resources Directorate
Logistics Research Division
Wright-Patterson Air Force Base

INTRODUCTION

I was recently amused by a cartoon depicting three cavemen staring up in confusion at a large diagram covering the wall of their cave and debating its worth. The diagram included a number of circles. Each was labeled with categories like, "Critical Hunting Review" and "Ambush Preparation," and was connected by arrows indicating important relationships. One caveman says, "I don't know. It seemed easier when we just went hunting." Another caveman replies, "I know, but Ogg assures me this will improve efficiency and keep us ahead of the Cro-Magnon down in the valley." The caption read, "Why Neanderthal Man Became Extinct."

This cartoon is not only amusing because it accurately depicts a very common situation, but it is also very instructive. We might ask ourselves some of the following questions: Why do we build models? Are they of any real use? If they are, to whom do they provide the most benefit? What benefits can the customer expect by dedicating resources towards model development? If these benefits are considered important enough, how do we ensure getting the most for our time and effort?

Hopefully, this paper will answer these questions. I will begin by discussing the purpose of modeling, especially as seen from the perspective of a customer. Then I will present three methods used in the development of Air Force systems to illustrate why they were selected, what limitations and benefits they provide, and how to derive the most benefit from their use.

THE PURPOSE OF MODELING

Like me, you have probably heard a number of reasons to justify the time, the energy, and the money required to build models. I am not referring to the term "model" as applied to traditional engineering disciplines, such as a set of mathematical expressions that "model" the behavior of real world systems. Rather, the term as used here refers to the diagrams, or descriptions generated by integration experts through the analysis, or design of a proposed system. All too often, we find ourselves caught between the system developer on one side who claims we can't possibly afford not to build the model, while our funding source wants to know why we are doing the exercise in the first place.

A simple description of the product development process illustrating where modeling fits is seen in Figure 1.

Figure 1 depicts the customers' perspective of the activities involved in developing a product and their relationships to one another. That product may be night vision goggles, an information system, or even a house.

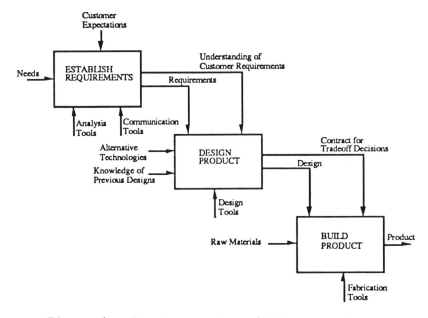

Figure 1. Develop product (IDEFo Model).

In IDEFo, the boxes represent activities that are performed. The arrows attached to the left side of boxes indicate inputs. Those attached to the right side of the boxes are outputs. Arrows coming into the top of a box represent controls or constraints relative to the performance of the activity. Those coming into the bottom of a box indicate the mechanisms by which the activity is accomplished.

We can see from the box labeled "Establish Requirements" that the customer provides input in the form of an expression of needs. Notice that this depiction refutes the popular notion that "...in the beginning was the requirement." Customers recognize problems in their environment that need to be solved, but for which the cause is often unknown. Needs, then, are conditions that must be satisfied. It is up to the product developer to establish the limiting constraints on conditions that must be satisfied, which are called requirements. For example, an architect designing a new home will take an expression of need for more space and lower heating bills by asking questions like, "How big is your lot size?" "What kind of additional space do you need? Storage space? Sleeping space?" "What is an acceptable range for heating bills?" and "How much money do you have to spend on this house?"

The results of this activity come in the form of two products: a clear set of requirements and a perception that the customer environment is understood well enough by the developer that the needs will be satisfied. This perception is a byproduct of the developer's attempt to isolate the causes of the problems through careful study of the environment where the symptoms occur. This leads to the rediscovery of the problems found by the customer, often resulting in the discovery of other problems.

The developer can use whatever analysis tools desired, as long as they promote lower costs and faster turnaround time. In fact, many of the actual products generated through the use of analysis tools may never be seen by the customer. Periodically throughout the analysis process, it is necessary to communicate to the customer what the developer is actually thinking. This amounts to going back to the customer and asking, "Is this what you mean?"

The second activity, "Design Product," will typically not proceed without a clear set of requirements, and a feeling of assurance from the customer that expectations will be satisfied. The product developer then uses whatever design tools will best help to satisfy the requirements. There may also be cases where communication tools are used by the developer to communicate with the customer throughout the design activity. An architect, for example, may simply generate a set of blueprints from the requirements without needing to consult further with the customer. Cases where it is necessary to involve the customer further will most likely occur when trade-off decisions have to be made. The developer must then demonstrate how competing design decisions will affect the cost of the end product. The trade-off decisions made at this point can then be captured either implicitly, such as by verbal agreement, or explicitly, through formal sign-off.

Once the design is accomplished, and trade-offs have been accepted by the customer, the actual building of the product can begin. The fabrication tools to be used in this process can range in sophistication from totally automatic generation of the product to strictly manual approaches.

There are two places where it should not be more obvious how models can be useful to the customer. Models built as part of the product development process should:

1. Instill a feeling of assurance that the developer understands the customer's environment and the conditions that must be satisfied to meet expectations.

2. Involve the customer in making trade-off decisions impacting the developer's ability to meet expectations and serving to document those decisions.

One important purpose for modeling, from the standpoint of the customer, is to satisfy these needs. A corollary to this assertion is that if the developer is building models that do not satisfy these roles, success in meeting the customer's needs is left largely to chance.

This should in no way diminish the importance of models to satisfy the needs of the product developer. Rather, these guidelines should be used to help discern what kind, and how much modeling is actually needed.

Now that the purpose of modeling, as it relates to the customer, has been explained, I can discuss some of my experiences in the use of the IDEF (Integrated Computer-Aided Manufacturing DEFinition) methods to perform modeling activities in support of product development. Three such methods will be described: IDEFo (the zero is subscript, indicating that this is the IDEF method of function modeling); IDEF1 (Information Requirements Modeling Method), and IDEF1x (Data Modeling Method).

FUNCTION MODELING USING IDEFo

IDEFo has been successfully used as both an analysis tool and as a communication tool in a number of application arenas. Referring back to Figure 1, this characterization indicates that IDEFo can be applied as a mechanism for performing the "Establish Requirements" activity. IDEFo is useful as an analysis tool, and it should be obvious by now that it also works very well as a communication tool.

The function modeling method, as IDEFo is called, was derived from a well established graphical language known as the Structured Analysis and Design Technique (SADT). The Air Force commissioned the developers of SADT to develop a function modeling method for analyzing and communicating the functional perspective of a system. Effective IDEFo models help to organize the analysis of a system, and promote good communication between the analyst and the customer. One other valuable feature of the IDEFo modeling method is its usefulness in establishing the scope of analysis, in particular for functional analysis, but also for other analyses which may be accomplished in the future from another perspective of the system under study. Thus, IDEFo models are often created as one of the first tasks of a system development effort.

IDEFo ORGANIZATION MECHANISMS

A number of organization strategies have been designed into the IDEFo methodology which, when properly used, lend tremendous expressive power and ease in communication. When improperly used, or when simply not understood, they yield models that are not only difficult to understand, but can make absurd declarations which appear well founded.

There are two main purposes for the organization strategies employed in IDEFo. For the modeler, they serve to focus work on one piece of the model at a time. They also establish clear boundary conditions within which to perform the analysis. For the customer, they allow rapid discovery and inspection of the system. They provide powerful browsing mechanisms for learning about the system under study as a whole and communicating the modeler's understanding of the system as well. I will first address purpose and viewpoint as an organization mechanism, then briefly focus on two other important organization mechanisms used in IDEFo: the hierarchical or top-down analysis approach to model development, and the notion of levels of abstraction.

To begin an IDEFo modeling activity, or an activity analysis, the modeler must first determine the purpose of the model, and from what viewpoint the activities will be defined. One purpose for using IDEFo might be to provide insight into overcoming known problems or identifying opportunities for consolidation. The viewpoint establishes how the reader will interpret the model, as well as how the modeler will constrain his idealization or abstraction of the activities that occur in the system under study.

Another mechanism in the development of IDEFo models is the notion of hierarchical decomposition of activities. Although IDEFo models are said to be developed using a hierarchical, or top-down approach, Doug Ross, the creator of SADT, is said to have often struggled with these terms. In fact, Mr. Ross proposed that this process might more accurately be characterized as an Outside-In approach. A box on an IDEFo model, after all, represents the boundaries we have drawn around some activity. We can look inside and see the break down of that activity into smaller activities, which together, comprise the box at the higher level.

This mechanism assists the analyst by providing completeness criteria in the form of the boundary conditions represented by the box being decomposed, with which to assess model quality. The customer will also find this organization mechanism useful for screening unnecessary complexity from view, until more in-depth understanding is sought by "looking inside the box" at its decomposition as seen in Figure 2.

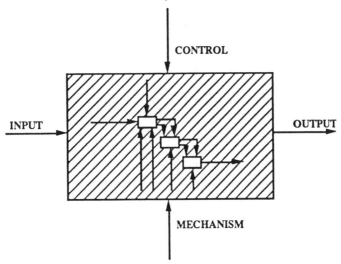

Figure 2. Looking Outside-in.

This screening may also be characterized as part of the abstraction mechanism used in IDEFo. One common misconception, however, is that levels of abstraction are only evidenced in the activities themselves as one moves between levels of the model. The arrows too, exhibit different levels of abstraction between model levels. In fact, achieving the correct balance between the level of abstraction associated with a box and the level associated with the arrows attached to the box is not always trivial. Let's suppose for example, the four mechanism arrows inside the bigger box in Figure 2 represent different types of tools used to accomplish the activities they are attached to. These four arrows could be "bundled" together into a more abstract perspective as one arrow labeled "tools." The big box in Figure 2 would have only one mechanism arrow for its level of abstraction. A far less elegant depiction would be where all four arrows appear at the more abstract and the more detailed levels of the model. As you can see, one way to tell whether the modeler has effectively used the information hiding constructs available in IDEFo is to count the number of arrows attached to the boxes at any level. If the model seems cluttered with arrows, it is likely that the level of abstraction used in bundling the arrows is not the same as that of the box.

Perhaps the least understood, or most frequently misapplied tool for screening unnecessary detail at a given level of abstraction is the notion of bundling and unbundling of arrows. It would be logically inconsistent, for example, to unbundle all but two of the four mechanism arrows in Figure 2 when each occurs at the same level of abstraction as component tools. Likewise, it would make little sense to unbundle and rebundle an arrow at the same level of abstraction. Unfortunately, the IDEFo literature does not adequately cover how to appropriately avoid logical inconsistencies introduced through incorrect use of information hiding constructs available through arrow bundling. The best approach to this dilemma is to build models using an automated support tool that enforces good practice. Otherwise, an IDEFo modeler can take years to learn how to recognize and avoid bundling problems.

SOME HELPFUL HINTS

There is lots of help available for those who want to be good IDEFo modelers. I will only attempt to offer a few of the more common hints to getting the most out of an IDEFo modeling exercise. Specifically, these guidelines will help you determine how well a model satisfied the purposes of modeling outlined above.

The most difficult thing to master in IDEFo is learning to maintain the same purpose and viewpoint as you move between levels of the model. What makes it difficult is that shifting viewpoints are difficult to recognize. A good rule of thumb is to look at the boundaries within which the modeler develops a decomposition and formulate questions like, "Does this activity fall within the scope of the higher-level activity?" and "Does this activity conform to the established viewpoint and purpose of the model?"

Another thing to look for are models which push the limits. For example, the discipline component of the IDEFo method establishes a rule that there should never be less than three, or more than six boxes to a decomposition. Likewise, there should never be more than six arrows on the side of a box. The tendency of most modelers is to draw six boxes first for a decomposition, and then come up with the names of the activities. What usually happens with this approach is that a seventh activity somehow becomes indispensible. Likewise, boxes begin to look like wiring diagrams for electronic components. This results from failure to organize the arrows themselves logically into bundles at different levels of abstraction, according to the level represented by the boxes to which they are attached. Inappropriate bundling, such as that done simply to abide by the established rules, may also occur. These kinds of errors become obvious when arrows seem arbitrarily grouped together.

Another common problem emerges when new conventions are introduced into the method. For example, some modelers will choose to establish a convention that inputs and outputs can only be data. In this way, they hope to ensure that inputs and outputs translate directly into information model elements. Intuitively, this approach would provide clear and unambiguous tracking of data needs across activities and clearly delineate the scope of their information models. However, with this approach, the modeler is forced to further change conventions by making mechanisms into resources assigned to an activity. For example, we could imagine an activity called "Fix Broken Airplane." With these conventions, there is only one place to input the broken airplane; as a resource assigned to the activity. There is, however, no place to show that what emerges is a fixed airplane, since outputs can only be data. It is far more efficient to use existing conventions, and then use inputs and outputs as candidates for what may be data to be examined and discriminated later.

The most useful tool for assessing the quality of an IDEFo model is to read it and see if it makes sense. Give yourself no more than two minutes per page. For a large model, this should translate into no more than two hours. If you can understand the environment that is modeled by the IDEFo representation in that time or less and feel like you can explain what goes on in that environment to someone else, it is highly likely that the model will be of significant value. Cases where models require two full days of careful study to fully comprehend are not good IDEFo models.

More sophisticated methods for assessing the quality and usefulness of IDEFo models are available to those who need them. But these simple rules of thumb will most likely be of equal, or greater value.

INFORMATION MODELING USING IDEF1

Again referring back to Figure 1, IDEF1 can be viewed as a tool for both analysis and communication in establishing requirements. In this case, however, IDEF1 is concerned with establishing the requirements for what information is or should be managed by your enterprise.

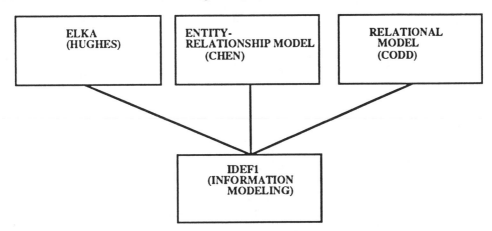

Figure 3. IDEF1 origins.

The information modeling method, as IDEF1 is called, derives its foundations from three primary sources: The Entity-Link-Key-Attribute (ELKA) method developed by Hughes, the Entity-Relationship (ER) method proposed by Peter Chen, and Codd's Relational Model. The original intent of IDEF1 was to capture what information is or should be managed about objects falling within the scope of an enterprise. The IDEF1 perspective of an information system is one which includes not only the automated component, or the computer, but also humans, filing cabinets, telephones, etc. Rather than a design method, IDEF1 is an analysis method to identify:

1. what information is collected, stored, and managed by the enterprise;
2. the rules governing the management of information;

3. logical relationships within the enterprise reflected in the information;
4. problems resulting from the lack of good information management.

The results of information analysis can be used by strategic and tactical planners within the enterprise to leverage their information assets to achieve competitive advantage. Part of their plans may include the design and implementation of automated systems, which can process more efficiently the information available to the enterprise. IDEF1 models provide the basis for those design decisions. IDEF1, then, is not used to design a database. It provides managers with the knowledge required to establish good information management policy.

IDEF1 BASIC CONCEPTS

There are two realms that are important to those using IDEF1 to determine information requirements. The first is the real world as we perceive it. This means people, places, things, ideas, etc. The second is information. Here, we find information images of those things found in the real world. An information image is, therefore, not the real world object, but only the information that we collect, store, and manage about real world objects.

Let's focus on the first realm. The term "Entity" is used in IDEF1 to describe real world people, places, things, or ideas, as illustrated in Figure 4. For example, the sales department in a company is an entity, as is an employee working in that department. Entities may have characteristic properties or "Attributes" associated with them, such as a name, age, gender, etc. Further, one entity may be involved in some kind of association with other entities, called a "Relation." The term "works for" describes the relation between an employee and his or her department.

Now we can turn to the information realm. An "Entity Class" is a collection, or class of information we keep about entities in the real world. Think of an Entity Class an an empty box for holding 3 x 5 cards. The box is labeled on the outside with a name describing what type of cards go in the box, and a template for the individual cards that will eventually go inside. The template is made up of names of attributes (characteristics or properties) of the objects under study. An "Attribute Class" is the set of attribute-value pairs formed by the name of the attribute found on the outside of the file box. This class also is made up of the values of that attribute for individual entity class members listed on the individual cards themselves. One or more attribute classes, allowing us to distinguish one card from another, or one member of an entity class from another, is called a "Key Class." A key class is indicated by placing it in the top left corner of the template and having it underlined. A "Relation Class" represents a reference made from one member of an entity class to another using the key class of the referenced entity class member.

SOME THINGS TO LOOK FOR

Some simple ways exist of inspecting an IDEF1 model to determine whether the modeler has your expression of need, and an understanding of your environment, and has successfully identified existing and future information

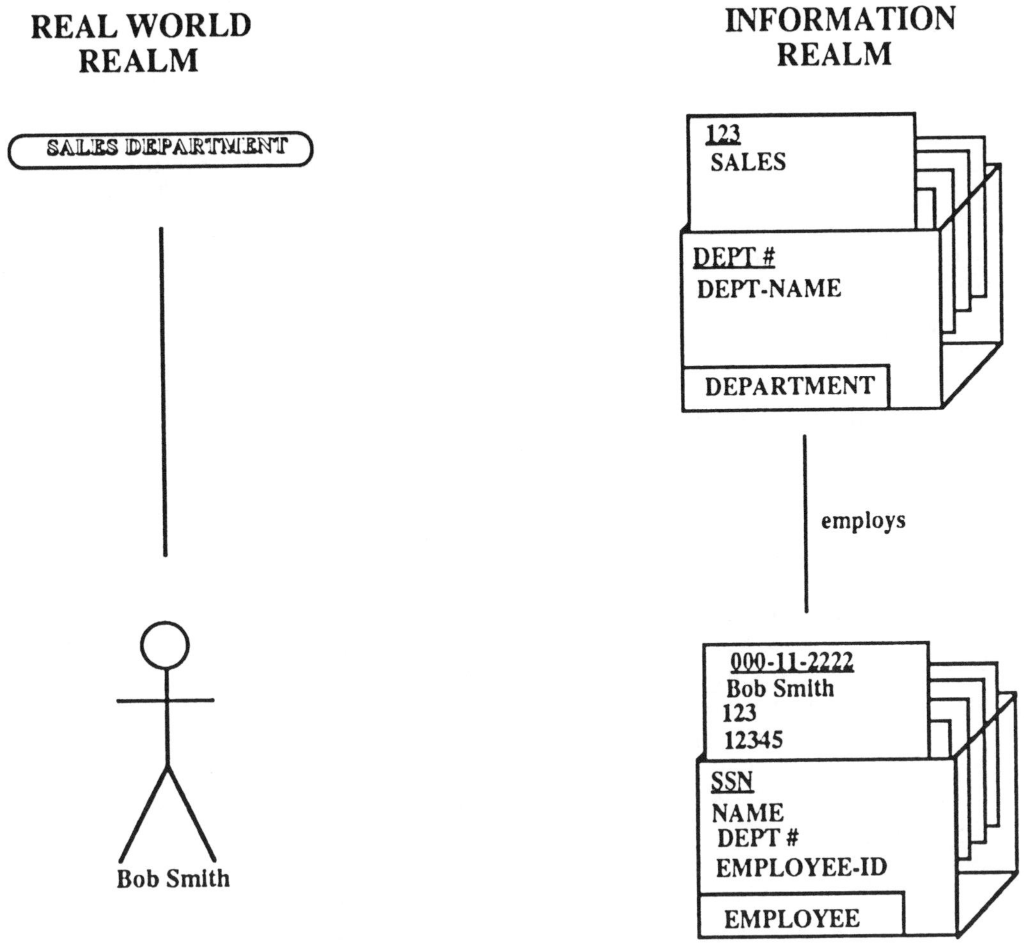

REAL WORLD REALM

SALES DEPARTMENT

Bob Smith

INFORMATION REALM

123
SALES

DEPT #
DEPT-NAME

DEPARTMENT

employs

000-11-2222
Bob Smith
123
12345

SSN
NAME
DEPT #
EMPLOYEE-ID

EMPLOYEE

Figure 4. IDEF1 basic concepts.

management requirements. These techniques assist the customer in determining if the IDEF1 modeler is modeling things in the real world realm or information realm.

First, examine the labels used to name the entity classes. If you notice plural rather than singular ends, for example, it is unlikely the model was created with the information realm in mind. A box labeled "Person" will have a better chance of conforming to the convention that the entity class represents--the set of information that is important to manage about a real world person--than one labeled "People." When the latter label presents itself, it is most likely that the modeler means a group of people with the same attributes.

Models which represent real world things have some boxes with no non-key attribute classes and some with whole lists. Those with no non-key attribute classes will typically be things like organizational units. Close inspection of boxes with long lists of attributes will often reveal that the box represents a form such as a purchase order, with the attributes being the actual fields on the form.

The third thing to look for is models with most of their boxes having only one or two lines attached. In this case, it is highly probable that the box with all the lines attached to it will be the box labeled "Person." Many of the lines attached to the box will indicate the roles that a person can assume. For example, we might read things like, "Person inspects Parts," "Person certifies Inspection," or "Person is married to Person."

It is natural for people using IDEF1 for the first time to model the real world in terms of the things they can easily see. Misapplication of IDEF1 may lead to huge models with little or no benefit, such as the one found on the cave of the Neanderthal tribe in the cartoon described earlier. With little direction, the novice modeler attempts to model everything observed in the real world. These models become difficult to manage as they grow. Soon, it becomes easy to lose sight of the purpose of modeling. Information models that model the real world realm rather than the information realm, provide virtually no insight into what information requirements are, or where more effective information management policies can improve competitive posture.

THINGS TO CONSIDER WHEN USING IDEF1

When using IDEF1, avoid naming the lines that connect IDEF1 boxes, or the relation classes. Looking again at Figure 4, it seems very intuitive to read the model as, "a department employs one or more employees." Is this the same as, "the information we keep about departments employs the information we keep about employees"? Obviously, the first reading confuses one into thinking that the file box is actually the real world object, and not the information that we keep about an object. Remember: the link between the boxes represents a reference from one file card to another, or information relationships, not real world relationships.

A sentence like, "A department employs one or more employees" is a natural language fact, or a business rule. Since business rules are important to know and manage, perhaps they should be tracked and managed in similar fashion to the way we track and manage source data items in IDEF1. The only exception would be that we may want to manage them separately as a source facts list. In the meantime, we can label the links between entity classes with things like L1, L2, etc. This would make it very clear that a relation class represents a function which, when applied to a member of one entity class, will return those entity class members involved in the information relationship.

Information relationships and real world relationships cannot be indiscriminately mixed without leading to confusion, paradox, and error.

For example, suppose we create an entity class labeled "System" and attach it to a link that originates and ends at the System Entity Class with the class label, "is comprised of." How do you interpret such a model?

Intuitively, it would seem very reasonable to interpret the box as representing a system-subsystem hierarchy. The highest level system may be something like an automobile which would be broken down into subsystems and subsystems, down to the individual component level. Using this approach, it appears that we could ascribe the same system-subsystem hierarchy to the information we keep about a vehicle to help us think about its organization more intuitively.

Can we now say that the information we keep about the starting system is comprised of information about systems falling below it in the system hierarchy? This wouldn't work because the information we keep about a car would include things like who owns it, what state it is registered in, what year it was manufactured, etc. None of that information has anything to do with the information we keep about subsystems like the braking system, which might include the type of front and rear brakes, the moisture content in the brake fluid, or the remaining thickness of brake pad. Obviously, we cannot assume that information relationships will behave the same way that real world relationships do.

The only correct interpretation for an IDEF1 model, which avoids unnecessary confusion, is that one member of the class of information we keep about an object can reference information about other objects through the referenced member's key class.

DATA MODELING USING IDEF1x

Once I have a thorough understanding of what my information requirements are, I can begin to make decisions about how to manage that information more effectively. One such decision might be to implement an automated system, requiring the selection of an appropriate design method. IDEF1x is one alternative for logical database design.

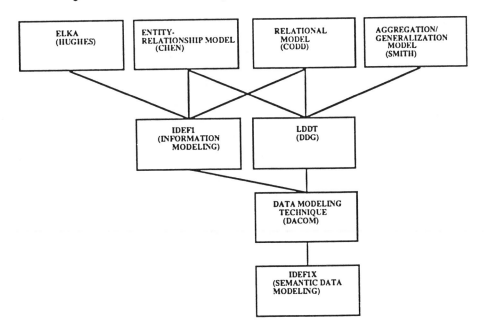

Figure 5. IDEF1x origins.

IDEF1x was influenced in its development by Chen's Entity Relationship (ER) model, Codd's Relational model, and Smith's Aggregation/Generalization model. These origins led to the development of the Logical Database Design Technique (LDDT) by the Database Design Group, Inc., which later became a commercial product of Dan Appleton Company (DACOM) known as the Data Modeling Technique (DMT). DAMOM then used this method and their experience with IDEF1 to create the IDEF Data Modeling Method, or IDEF1x, which is pictured in Figure 5.

Referring back to Figure 1, we can say that IDEF1x is intended as a mechanism for accomplishing the "Design Product" activity. Because it is a design method, IDEF1x is not targeted to serve as a requirements analysis tool. IDEF1x picks up at the data design point, after the information requirements are known, and once the decision to implement in a relational database has been made. Hence, the IDEF1x perspective of the information system is restricted to what will actually reside in a relational database. If your target system is something other than a relational system, like an object-oriented system, IDEF1x may not be the best method of choice.

There are several reasons why IDEF1x is not well suited for anything other than relational system implementations. IDEF1x requires, for example, that the modeler designate a key class to distinguish one entity from another, whereas object-oriented systems do not. Further, in those situations where more than one attribute or set of attributes will serve equally well for individuating IDEF1x entities, the modeler must designate one as the primary key and list all others as alternate keys. The explicit labeling of foreign keys, which are attributes shared by one entity but which serve as the key attribute in another entity, is also required. Modeling constructs such as these, clearly indicate specific intent to accommodate logical design for relational systems implementations. Hence, the results of an IDEF1x design activity are intended to be used by those involved in taking the "blueprint" for an information system, or the logical database design, and implement that design in a relational database.

INDEF1x BASIC CONCEPTS

Since the terminology used in the IDEF1x method is similar to that used by the IDEF1 method, some further definition of terms is necessary to avoid confusion. Fundamental differences also exist in the theoretical foundations and concepts employed by the two methods. An "Entity" in IDEF1x refers to a collection, or set, of like instances (persons, places, things, or events) that can be individually distinguished from one another. A box in IDEF1x represents a set of things in the real world realm. An "Attribute" is a characteristic associated with each individual member of the set, called an "Entity Instance." Referring to Figure 6, Bob Smith and the Sales Department are both entity instances. DEPARTMENT is the collection of departments; EMPLOYEE is the collection of people that are employed by individual departments. Department Name, Department Number, and Department Location might be attributes of the Department Entity. The "Relationship" that exists between individual members of these sets is given a name. In this case, we can correctly read that "A department employs one or more employees."

Notice here that IDEF1x actually attempts to model the real world through the use of classification structures. Another classification structure included in the IDEF1x method, not present in IDEF1, is the generalization/specialization construct. This construct is an attempt to model "kinds of" things, whereas the boxes, or entities, attempt to model "types of" things. These "categorization relationships" as they are often called, represent mutually exclusive subsets of a generic entity or set. In other words, subsets that emerge from the superset cannot have common members. One example of how this might be used is to state that given a generic entity called PERSON, we can create two subsets which represent a complete set of categories of people, namely, MALE and FEMALE. No member of the MALE set can be a member of the FEMALE set, and vice versa. The unique identifier for a member of the male set is, by definition, the same as that for a member of the generic entity. The same holds true for the female category entity. The general attributes that apply to all members of the entity PERSON are listed in the generic entity. The specialization attribute, gender in this case, is listed in the category entity.

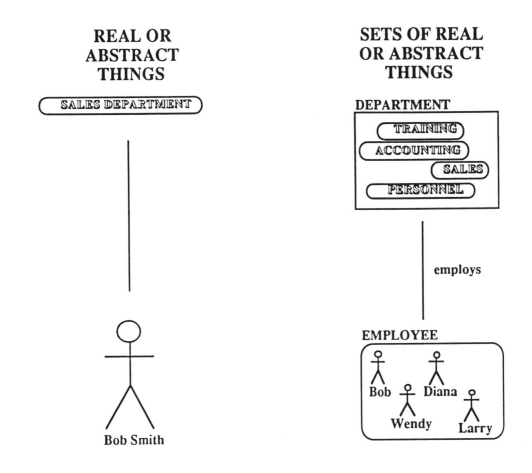

REAL OR ABSTRACT THINGS

SETS OF REAL OR ABSTRACT THINGS

Figure 6. IDEF1x concepts.

THINGS TO CONSIDER WHEN USING IDEF1x

As you may have already noticed, the underlying concepts associated with the IDEF1x method are intended to model general knowledge structures, or natural language facts about real world things, people, places, events, etc. This is much different than the goal of IDEF1. IDEF1x goes a step further by attempting to design logical data structures associated with these sets of real world things. That is meant to be accomplished through the definition of attributes and relationships.

If we consider again the examples discussed in the section on IDEF1, we can see that the intent of IDEF1x is to model things like, "A department employs one or more employees" or "A system is comprised of one or more subsystems." Since we have already established that real world relationships between objects do not behave the same way that information relationships do, we should ask ourselves if this approach is useful for designing logical data structures. The reader may want to address this question by reconsidering the example of the model representing hierarchies of systems.

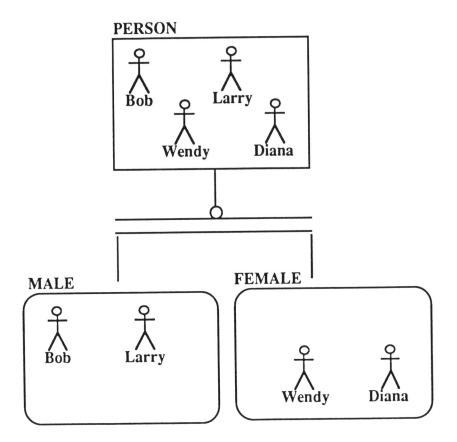

Figure 7. IDEF1x generalization/specialization construct.

Another common example is found in an entity called Technical Order
Improvement that actually appeared in a model delivered to the Air Force,
and meant to establish the specification for a logical database design. I
should have put the word "Form" on the end of the entity name, since the
attributes listed in the entity actually correspond to individual fields on
an Air Force form called the AFTO 22. The entity here is used to model an
artifact, the form itself, hence a laundry list of attributes.

The first question we might ask ourselves is, "Is this a reasonable model of
the set of forms called AFTO 22s?" This entity certainly presents an
accurate model of the form and may, therefore, be fully justified. The next
question we should ask is, "Has any design activity been facilitated through
modeling the form in this manner?" In other words, "Does this kind of
modeling assist me in moving from a set of information requirements to the
logical design of relational tables, and from which I can make trade-off
decisions?" Does it make sense to create a table, or relation in a
relational database that corresponds directly with information on existing
forms? This approach may have little or no impact for small implementations
where only a limited set of artifacts have to be modeled. However, for
large-scale implementations, the variety of different form types alone would
easily employ an army of modelers doing inventory of the forms and the
fields that go on the forms.

Based on these examples, it should be obvious that this approach to logical
database design can lead to confusion and/or poor design. So what if we
attempted to establish the convention that an entity must represent
information kept about real world objects rather than the objects
themselves? By doing this, we would have little more than a syntactic

variant of IDEF1 with one exception: the generalization/specialization construct. This, too, could lead to confusion if used without exercising great care. For example, is it true that the set of information we keep about people is divided into two types: male information and female information?

Perhaps we could use IDEF1x as simply a method for modeling just the objects themselves, not the structure of the data associated with them. Models of this type are often called concept models. Again, we run into difficulty with the generalization/specialization construct, since in the real world, we talk about overlapping rather than mutually disjoint sets. Since IDEF1x does not allow for categorization of entities with mutual members, IDEF1x could not serve well as a concept modeling language. For example, a model of the kinds of employees that we might find in a company might include managers, engineers, designers, and secretaries, among others. A generalization entity in this case would be EMPLOYEE, and the specialization entities would be those listed above. The interpretation of this structure would imply that engineers cannot be designers, nor can they be managers.

Clearly, we need methods that can effectively capture what we know about the real world and the relationships that exist between people, places, events, etc. The popularity of several methods which have built-in constructs for modeling these things is only a portion of the body of evidence supporting the reality of this need. We obviously need methods that capture existing and anticipated information management requirements. Unfortunately, there are a number of commonly used modeling languages, which fail to maintain an unambiguous distinction between the two realms.

CONCLUSION

Going back to Figure 1, I have talked about methods used to accomplish the analysis and design of a system. IDEFo serves particularly well as a tool that can do functional analysis and communicates the activities embodied in a given system or environment. IDEF1 also serves well as an information requirements method and a communication device for conveying what information is or should be managed by an enterprise. Some care is required, however, in the use of relation class names to ensure that the full power of the method is taken advantage of without introducing unnecessary confusion. IDEF1x has been introduced together with its intended use as a design method.

Each of these methods represent attempts to address the increasing need to manage complexity by dividing up the systems we must analyze, design, and develop into discrete, manageable chunks. An appropriately designed method serves to raise the level of performance of the novice to a level comparable to that of an expert. Hence, methods themselves embody knowledge of good practice for a given analysis, design, or fabrication activity.

For the customer, floor plans and artist's renderings are just as important as the final blueprints. It is important that the methods developer constantly re-evaluate how well individual methods serve the needs of the modeler and the end customer. Practitioners must become sufficiently familiar with the basic theory behind the methods to ensure their appropriate selection and use.

Just as the original IDEF methods were targeted at managing the complexity associated with evolution towards large scale integration in the manufacturing environment, new challenges will continue to emerge as those visions extend to integration across traditional boundaries as well. Large-scale integration between engineering, manufacturing, and support activities will be exciting and challenging, particularly to the methods engineer. These people will be responsible for encapsulating in easily usable form, the basic theory and body of experience associated with the analysis, design, and realization of tomorrow's integrated environments.

A Case Study: Integrated Methods and Tools for Optimization

by Raymond Vanderbok and
John A. Sauter
Industrial Technology Institute

INTRODUCTION

Modern manufacturing is becoming increasingly complex. New processes, greater product variety, higher quality standards, and shorter product life cycles all contribute to growing demands for more sophisticated manufacturing systems. Unfortunately, our ability to design and build systems of greater flexibility and sophistication is not keeping pace with this increasing demand. Frequently, flexible manufacturing is delivered late, is over budget, and rarely meets expectations.

If American manufacturing is to remain competitive in tomorrow's markets, this situation must change. Traditional strengths in technology need to be leveraged to compete in a global marketplace. The Industrial Technology Institute (ITI) is addressing these problems in manufacturing design with new methods and tools. These tools have been used in the design of several actual manufacturing systems. This case study investigates applications where these new techniques have been used to address shortcomings in the current design process. Two applications are described along with the experiences gained, and the improvements observed.

PROBLEMS WITH ADVANCED MANUFACTURING SYSTEMS

There has been little research into the reasons for the failure of manufacturing systems. However, studies confirm our own experience that the number one problem is software (1). Alting, in his paper (2), classified the software problem into three areas:

1. Requirements of the system, software, and equipment are not correct. Requirements are difficult to capture in a way that customers and software engineers can use. Requirements tend to change during the life cycle of the system, having impacts on the system life cycle.

2. Software development does not satisfy requirements. The lack of good requirements models for manufacturing systems makes it difficult for software engineers to develop software that meets the customer's expectations.

3. Interfacing among software, computers, machines, devices, and systems is inefficient. This is a significant factor in the excessively long integration and debugging associated with most advanced automation projects.

PROMISING APPROACHES

Several organizations are working to improve the way advanced manufacturing systems are being designed. A variety of techniques are being applied to various aspects of the problem.

Programs like STARS (3) and other research initiatives are investigating the general problem of software development. Object-oriented programming and rapid prototyping are emerging as two of the most promising technologies for improving software development in the 1990s. Application of these technologies to the design of manufacturing systems will be important elements in any effort to improve factory systems.

Physical prototypes have been used in product design for many years. Developing a scale model of a complex part helps resolve many errors in product design. Building a working model of a manufacturing system, however, is expensive and takes months to complete. Unlike products, manufacturing systems are typically one-of-a-kind systems, offering no opportunity to recover the investment in the prototype through sales of repeated copies of the system.

Computer models provide another method for prototyping by simulating physical behavior. To be a useful replacement for a scale model, the simulation must be detailed enough to exhibit all the dynamic characteristics that the scale model would have and more. The information and experience gained through building and executing the simulation model are at least equivalent to that of building and testing a scale model of the system.

Software modeling is a well-accepted practice in software engineering to describe the structure of data and control. Software models cannot model hardware. In traditional data processing applications, this is not a problem. In manufacturing control, the primary function of the software is not to manipulate data, but to control and interact directly with the mechanical equipment. Models are needed which allow direct observation of the behavior of software interactions with the physical hardware of the system.

Clearly, there is a need for better methods and tools for rapid development of detailed models of hardware and software which can be simulated and animated. Ideally, these models should not only answer questions about the system, but serve as specifications allowing rapid implementation of the final system.

XSPec (TM) AND XFaST (TM) METHODS AND TOOLS FOR SYSTEM OPTIMIZATION

ITI has developed a system design methodology called XSpec (TM) (eXecutable Specification) (4) which is based on object-oriented principles and rapid prototyping. A prototype tool framework called XFaST (TM) (eXecutable

FActory Simulation Tool) (5) has been developed to support the methodology. This framework integrates three commercially available tools: SIMAN (TM), FLEXIS (TM), and ROBCAD (TM). These tools are used to develop different kinds of simulation models (discrete event, kinematic or control) which are then executed together and synchronized through XFaST (TM). XFaST (TM) can be extended to integrate other simulation tools as needed.

In XSpec (TM), a system to be modeled is partitioned along the lines of the objects (the data and equipment) which make up the system. The basic modeling object in XSpec (TM) is called an element. An element is used to represent the behavior of some part of the system. The element is an executable model. It also serves as a specification for that part of the system. In XSpec (TM), these elements can be plugged together with other elements and executed on a computer.

There are two kinds of elements in XSpec (TM): physical elements and control elements. Physical elements represent factory devices or subsystems (the hardware) which respond to some controlling agent. Control elements represent those controlling agents (the software) in the system. An example of a physical element is a section of conveyor track. It would include the motors to drive the conveyor and any sensors used. A control element represents the software that reads the conveyor's sensors and controls its motor.

These elements are naturally grouped into a logical construct called a component. A component groups related control and physical elements together. These elements represent the hardware and software which make up a distinct piece of factory equipment. This grouping allows a natural approach to system decomposition. Figure 1 shows an example of the notation used in XSpec (TM) to describe a component.

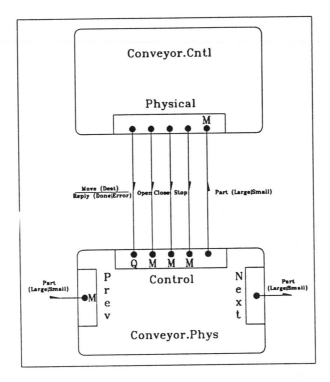

Figure 1. Example of an XSpec conveyor component.

Like objects in object-oriented systems, components exist in a class hierarchy. Manufacturing devices naturally fall into a class hierarchy. For example, you might have a class for all material transport devices. Under this class might be several subclasses, such as robots, AGVs, and conveyors. Conveyors could be further divided into accumulating and nonaccumulating conveyors, and so on. The class hierarchy provides a convenient mechanism for grouping components and deriving new components when they are needed. This facilitates the reuse of design components. XSpec (TM) uses the term message to describe all interactions between elements. A message represents the flow of information and material in the system. In an actual system, information flow can be accomplished through many different means: digital I/O, serial lines, local area networks, common memory, etc. XSpec (TM) can model each of these modes of communication through messages. This keeps the details of information flow out of the design while still giving the designer the ability to accurately model a particular system. Messages also allow the specification to be executed through standard computer techniques. The element models are created using traditional simulation tools in the XFaST (TM) environment. Each element can be modeled on the appropriate tool. Physical elements are usually modeled with SIMAN (TM), and animated with Cinema. Robot elements can be modeled with ROBCAD (TM), a kinematic modeler. Control elements are normally modeled with Savoir's FLEXIS (TM), using a language called Grafcet (also known as Structured Function Charts). These models can then be executed together to observe the behavior of the entire system. When tested, the models serve as specifications for software or mechanical implementations.

CASE STUDY #1:   MATERIAL TRANSPORT SYSTEM

This section describes how the XSpec (TM) methodology and the XFaST (TM) tool were used in the design of a highly modular Material Transportation System (MTS). This example demonstrates how integrated tools can create robust designs that promote reuse. The complex MTS was decomposed into an interconnected set of components that include physical hardware and control strategies. Design alternatives were evaluated by including simulation as an integral part of the design process. Design errors were identified during simulation. System performance was validated through simulation. The system, as represented on the tools, form a specification for implementation. The Grafcet used for control during simulation can be translated to the target controller. The definition used to model the physical aspects of the system can become specifications for the mechanical builders.

The resulting design accommodates a variety of conveyance technologies. The user can modify process routing and the number of work stations through simple data table changes, rather than software changes.

APPLICATION

This MTS application was designed for a large manufacturer of electro-mechanical assemblies. To reduce the cost associated with frequent model changeover, the company desired an adaptive manufacturing line that could be quickly reconfigured to new product requirements. The intent was to provide a reusable design that was appropriate for low-volume (1,500 parts per day) production. The system was required to accommodate a mix of automated and manual process operations. The physical layout and control strategy needed to be flexible, so work stations could be added, deleted, or moved at low cost.

Production required that process routing be flexible. As the process plan for the part changed, the system had to adapt easily. The system then needed to be flexible in assigning process operations to work stations and flexible in routing material between work stations.

The process plan required parts to visit a work station a number of times through the build. A given model of the part was run for about four hours before the system was changed over to a new model. Expected work station downtime was included in the design.

Because of the complexity of multiple visitation, dynamic pallet routing, and many resource constraints, the company realized the likelihood that the control method would directly affect system performance. The rationale for integrated simulation using XFaST (TM) was to evaluate system performance through simulation using the actual control strategy.

This project was used to evaluate the concept of modular material transport components including both the physical and control aspects.

Figure 2. Flexible material transport layout.

SYSTEM DESCRIPTION

The hardware selected used a dual belt accumulating conveyor with pallets. The hardware layout is shown in Figure 2. Each pallet carries four parts. The selected layout had an inside transport, and a holding loop. The outside lanes had work stations attached at standard intervals. Pallets could move between the inner loop and outside lanes. Pallets can also be passed directly from one work station to another. The hardware configuration also allowed transfers across the inner loop.

The system layout was modular and could be easily expanded. The unit of modularity, seen in Figure 3, included the track related to the work station, the associated section of the inner loop, and two transfer devices.

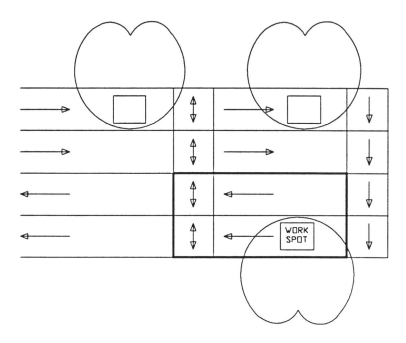

Figure 3. Material transport module.

CONTROL STRATEGY

The control for the MTS existed at two levels. First, device control was concerned with the actual logic that routes the pallet along a selected path. Second, routing strategy determined the routes which were to be followed. Device control involved sensor/actuator signals, control element interactions, and scope of control. Scope of control refers to the design decisions related to how parts of the physical system are assigned to be under the control of specific software modules. Routing strategy involved path selection and work station buffering policy. Simulation was used to select design alternatives and refine the behavior of both device control and routing strategies.

For device control, a strategy called Wheels (6) was used. Wheels considers all track-based Material Transportation Systems to be made up of interconnected Material Transport Components (MTCs). As shown in Figure 4, MTCs have convergent sections, storage sections, and divergent sections. All sections are optional. If the MTC has storage, it must have some method to stop the flow of material, either turning off a conveyor or raising a stop.

The material handling part of work stations are like MTCs with the addition of a stop at the work spot. For this application, each modular unit had one work station and three MTCs associated with it. The scope of control was very direct. The work station control mapped onto the work station track section. The related section of the inner loop and two transfer devices were each controlled by an MTC. With Wheels, all work station and MTC control elements were identical.

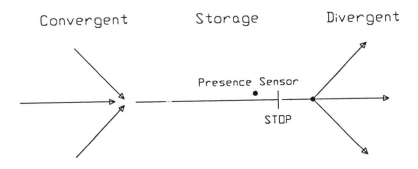

Figure 4. Material transport component.

MTS COMPONENTS

This application required three classes of XSpec (TM) components. First, is
the PalletTable component that provides routing information for the system.
Second, is the MTC. The MTC component includes the control and physical
hardware design related to material movement through a section of track.
Finally, the work station component is associated with the movement of
material through a work station. The work station component is similar to
the MTC component, but includes all the functionality of the work spot.

These MTS components included control and physical elements. The control
element was a representation of the control strategy. For this application,
the control strategy was implemented in Grafcet. The physical element was a
representation of the track design. For this application, the physical
element was modeled using SIMAN (TM) and animated using Cinema.
Importantly, these design constructs can be executed using XFaST (TM) to
evaluate the behavior of the system.

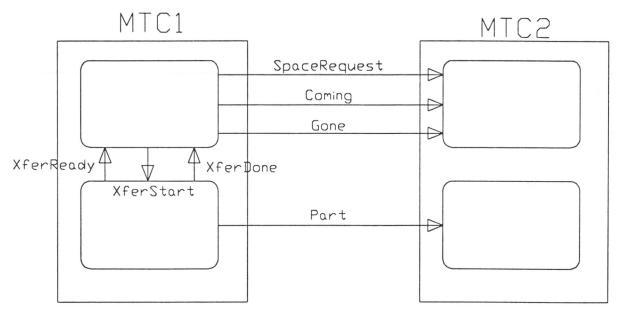

Figure 5. Component interaction.

The component diagram for two MTC components is shown in Figure 5. The
following describes the interaction between elements when a pallet is
transferred between two MTCs: when a pallet reaches the end of the track
section in MTC1_Physical, a sensor indication "XFerReady" is sent to its
control element. The MTC1_Control element then determines that the next

location the pallet should visit is MTC1. MTC1_Control then requests MTC2_Control for "SpaceRequest." When the requested space is granted, MTC1_Physical signals MTC1_Control that the transfer has been completed using a "XferDone" message. MTC1_Control then finishes the interaction by informing MTC2_Control "Gone".

For this system, one PalletTable, 34 MTCs and 10 work station components were used. This high level of duplication allows the designer to take advantage of the economy of scale. More effort could be applied to improving the performance of components because the benefits applied across the entire system. The description of the system is reduced to a description of how the components interconnect.

Components helped constrain the design to promote good system design practice. For example, when this system was first simulated, a problem was found relating to how MTCs negotiate for resources. The easy fix would have made future changes more difficult. Like structured programming, XSpec (TM) design rules helped promote good practice.

ROUTING STRATEGY EVALUATION

XFaST (TM) was used to evaluate routing strategies. This type of evaluation is normally done using stand alone discrete event simulation, but can be done within XFaST (TM). By evaluating the routing strategies within the system design environment, the design engineer can has greater confidence that the interactions between devices and control have been faithfully modeled. The designer can then select alternatives with lower risk. This was a good point in the design process to make trade-offs associated with mechanical versus control complexity.

The use of XFaST (TM) for this application helped to select a routing strategy and identify requirements that were initially missed. The initial routing strategy used the crossovers in the inner loop to form small loops that acted as virtual buffers close to individual work stations. Simulation with XFaST (TM) showed that this strategy could cause gridlock. The problem occurred when normal pallet traffic and buffer pallets filled the loop to absolute capacity. When a loop is filled to capacity, the pallets can no longer move between the MTCs along the loop. The control strategy requires that there be space for at least one pallet before allowing a pallet to enter a new MTC along the loop. The proposed solution was to use the whole of the inner loop as a shared buffer for all work stations. The inner loop could hold all pallets, so the capacity limit could never be hit.

The proposed change was tested, and again fell into gridlock. This time, buffered pallets on the loop combined with normal pallet traffic using the crossovers to over fill a small section of the inner loop. Two solutions were proposed. First, a new component could be added to the system to detect gridlock and reroute pallets to correct the problem. The alternative was to change the PalletTable and not use the crossovers. This alternative was selected favoring operational simplicity at the cost of slightly longer path length and associated work in process.

SYSTEM STABILITY

When this system was simulated, it was found to have production rates and work station utilizations that had large variations. The normal strategy for simulation is to allow the system to run long enough to reach a steady state. The production environment for this production line, however, included product changes every four hours. The system may never reach a steady state condition. In most cases, normal production will occur while the system is under startup dynamics. Through XFaST (TM) simulation, the dynamic behavior of the system was found to be greatly affected by the initial conditions. This understanding helped identify a missed requirement for the PalletTable component. System performance would be improved if the PalletTable component attempted to distribute pallets across the process plan. Localized overloading can happen within the system when many of the pallets are at the same point in the process plan and destined for the same work station.

SIMULATION PERFORMANCE

For simulation to provide insight into the design optimization, the simulation tool must facilitate rapid prototyping. The simulation provides a means of visualizing the system behavior, but it is equally important to be able to modify, adapt, and even replace parts of the system at low cost. XFaST (TM) allows control design engineers to make design changes using a tool made for expressing control strategies. This allowed control engineers to deal with the system from a representation they were familiar with. They could use the debugging aids that they felt were valuable in tracking down a problem. It would have been much more difficult for them to find a control design error from within a traditional stand alone discrete event simulation.

IMPLEMENTATION PATH

Because changes to the design were made to an executable specification, all aspects of the system included in simulation can be faithfully translated into the implementation.

The control elements in the XFaST (TM) model can be directly translated to the language of the target controller(s). In this application, the work station control element may execute in a robot controller. One work station control element may be assigned to execute in each robot. The design also provides the function of defining all interfaces between the robot and the MTS. A PLC may be used to interface to all manual work stations and control the rest of the MTCs and PalletTable. Implementation would require additional levels of detail, not included in simulation. This would include things like handling motor overloads and diagnostics.

The physical elements of the XFaST (TM) model can be used by system builders to define layout, track speeds, sensor locations, and I/O.

RESULTS

The company desired a modular system that was easily reconfigured. ITI verified that the concept of Wheels could meet production requirements through simulation on XFaST (TM).

The system designed used a few standard components that are replicated through out the system. This high degree of duplication within the system will facilitate system maintenance and continuous improvement. The people on the plant floor only need to understand the way material moves through an MTC once. That method was used for all movement within the MTS. What is learned on one part of the system can be applied to improve the entire system.

This MTS can be easily reconfigured and reused for other production lines. The modular nature of the hardware/software components reduces the effort to add or delete work stations or track sections. All that was needed was to define how the standard components interconnect. Simple data tables within an element identify neighboring elements. For example, to add a new section of track, the physical hardware and control code was copied. Then the data tables were modified for only the affected elements. Routing could also be changed without software. Data tables described the process plan for each product type and the assignment of process steps to specific work stations.

The modularity of Wheels and the design rules of XSpec (TM) have resulted in a system that can also grow vertically. Dynamic routing control or product scheduling could easily be layered on top of Wheels.

The company has reduced implementation risk through a comprehensive design process. Using this project, they were able to identify and correct problems at design time, when the impact on cost and schedule is minimized. The XFaST (TM) model can continue through the operations phase of the system lifecycle as a testbed for continuous improvement.

CASE STUDY #2:   GEAR WELD LINE

APPLICATION

A major automotive corporation asked ITI to assist them and their vendors in designing a transmission gear weld line. The vendors were located in different parts of the world. The line could not be integrated until all the pieces were assembled at the plant site. The company was interested in the unique capabilities of the XFaST (TM) tools in identifying design errors before system integration. Since there would be limited time for correcting problems after the line was assembled for the first time, XFaST (TM) could potentially eliminate expensive delays in system startup.

SYSTEM DESCRIPTION

The gear weld line consists of identical sides which share load and unload robots. Each side can handle 18 different part types with some changes in tooling. The cycle times for each part vary, so the two sides flow asynchronously.

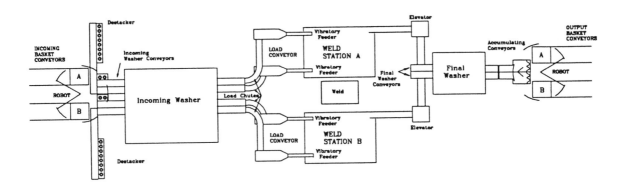

Figure 6. Gear weld line layout.

Figure 6 shows the layout of the line. Gears arrive at the robot in baskets on a conveyor. The robot unloads the gears onto the appropriate washer conveyor and coordinates the hub destacker which places the matching hubs on the lane next to the gears. After leaving the washer, the gears and hubs drop down gravity chutes onto separate load conveyors which lead to vibratory feeders. When the vibratory feeder fills, the load and washer conveyors are turned off. This controls the flow of parts into the weld cell.

Once the parts are welded, the single assembly is transferred via chutes and an elevator onto a final washer conveyor. After final wash, the parts are placed onto an accumulating conveyor where they are positioned for robotic pickup. The output robot loads the welded gears into a basket which is conveyed out of the cell. The weld station was designed to cycle parts fast enough to meet the company's production requirements. This placed a requirement on the material conveyance strategies to never starve the weld station for parts or block its ability to release finished parts. A traditional SIMAN (TM) simulation model was made of the entire cell to determine whether it would be able to meet production requirements. This model showed that the cell as designed would be able to meet and exceed those requirements.

MODELING THE LINE ON XFaST (TM)

ITI was asked to assist in designing the gear weld line. More specifically, the company needed:

* robotic strategies for loading and unloading parts from the line,
* working robot code for the GMF robots that were selected,
* conceptual design and requirements model for the hub destacker and the accumulating conveyor which suppliers could use for developing those components, and
* verification of the strategies being employed in conveying parts from the initial through the final wash.

Traditional timing studies and a kinematic modeler determined if a single robot could meet the line rates required. These studies showed a single robot was not fast enough, but there was not enough room to add a second robot. A dual gripper was designed which allowed the robot to move two parts at a time. With the dual gripper, it appeared that the robot would be able to meet the required line rates.

Next, the line was partitioned into its components. This layout is illustrated in Figure 7. There was a one-to-one mapping between the components and the different parts of the line. This made the partitioning process much easier than the traditional functional partitioning performed

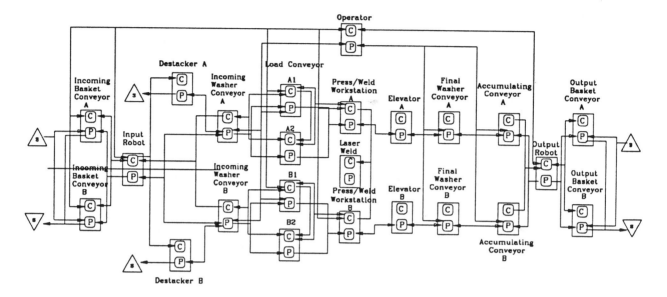

Figure 7. Gear weld component diagrams.

in structured analysis. Interestingly, the operator was modeled as a component in this diagram. This shows the interaction between humans and machines in semiautomatic processes.

Once the system was partitioned, the library of available components was scanned for applicable elements. Since our "library" only consisted of the three elements produced for the flexible material transport line, there was little to choose from. However, it was determined that the MTC component could be used after some modifications.

MTCs are general purpose material transport components that include converging and diverging switches. In this application, all the conveyor segments were much simpler; parts entered at a single point, and exited at a single point. For this application a simplified conveyor component was developed called a SISO conveyor (for single input, single output). Actually, two subclasses were used for the accumulating and nonaccumulating conveyors in the system.

Accumulating conveyors control the flow of parts by collecting parts behind some kind of stop. Nonaccumulating conveyors do not allow parts to shift with respect to the conveyor belt or chain. Flow can only be controlled by stopping the conveyor belt. For this application, it turned out that 16 of the 24 components (67%) could be modeled as one of these two classes of SISO conveyors. Five more classes were needed to define the remaining eight components.

The Savoir FLEXIS (TM) tool was used to define the control strategies with the Grafcet language. This graphical language allows complex sequences, expressions, or parallel behavior to be expressed easily and clearly. The robots were initially modeled with ROBCAD (TM). Once the robot timings were determined, ROBCAD (TM) was replaced with a simpler SIMAN (TM) model for the robots. The rest of the physical elements were modeled on the SIMAN (TM) tool, and animated with Cinema.

PROBLEMS IDENTIFIED DURING DESIGN

Once the components were defined and the appropriate control strategies modeled, all the system elements were simulated. The first runs of the model had a problem with parts colliding on the load conveyor. An investigation quickly revealed the cause.

The vendor had chosen a simple strategy for controlling the flow of parts into the weld station. When the vibratory feeder was filled, the load and washer conveyor were turned off simultaneously. This strategy would work if parts came down the conveyor at approximately even intervals (with roughly 10 cm between parts). Unfortunately, the dual gripper design caused pairs of parts to travel down the conveyor only 1 cm apart. If the two conveyors are turned off just when the second part of a pair begins to fall down the gravity chute, it will collide with the first part, which has not yet cleared the landing zone. Figure 8 shows the parts just when the conveyors have been turned off. The part at the top of the chute will collide with the part at the bottom.

The solution appeared to be simple; place a sensor in the landing zone to detect when the zone is clear of parts. The washer conveyor is turned off only when there is a part ready to drop, and there is no space at the bottom of the chute.

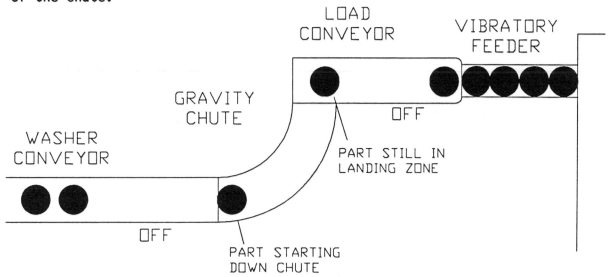

Figure 8. Part collision on load conveyor.

Other simple design errors were also detected. Because the designs were executed and tested with a variety of scenarios, it was possible to find many errors early in the life cycle. For instance, the original robot design had an error that caused it to enter into an infinite loop when a certain combination of events happened. Static walkthroughs, and careful designs were not able to detect the error. It was found after only a few hours of execution on the tools.

PERFORMANCE PROBLEMS

In designing robot strategies, there was concern for the ability of the robot to load parts fast enough to keep from starving the weld station. Even with the dual grippers, there appeared to be a potential performance problem. Since the two sides are asynchronous, the robot did not follow a fixed sequence. It did not appear to have time to wait at the basket until there was space on the conveyor before it committed to picking up the parts. A holding stand was added next to the washer conveyor. If the washer conveyor was turned off while the robot was waiting for space on the conveyor, it could drop the parts at the holding stand to service the other side. Later, when the washer conveyor was turned on again, and space became available, the robot would move the parts from the holding stand to the washer conveyor.

When the model was first run, a problem was immediately observed in the behavior of the robot. It was anticipated that the robot would occasionally have to use the holding stand. It was difficult to foresee, however, that the design of the vibratory feeder, 20 feet downstream from the robot, would drastically impact our robot strategies. When the vibratory feeder filled, the load and washer conveyors were turned off simultaneously. When there was space for another part, the conveyors were turned back on. To keep the weld station from being starved for parts, the input part flow was designed to continually fill the feeder. The model showed that this resulted in the load and washer conveyors being turned on and off for every part. The effect on the robot was disastrous. The robot got into a synchronous condition with the washer conveyor; every time the robot moved to the washer conveyor, the conveyor would shut off, forcing the robot to drop the parts onto the holding stand. This resulted in the robot making several extra moves for every part, reducing the part loading rate to 79% of that required.

A complete redesign of the robot strategy was initiated. A more careful timing study could be done with the integrated model, since it was possible to observe the complex interactions of the robot, conveyors, weld station, and destacker with different possible part mixes. A completely different strategy was designed and prototyped on the tool within two days. The holding stand was eliminated, and the complexity of the code was reduced 40% while increasing the overall performance of the robot by 50%. When the mechanical engineers were informed that the holding stand had been eliminated, they were quite relieved. It was turning out to be a significant problem to design a stand which would accurately hold parts with different outside and inside diameters in a fixed point for easy robotic pickup.

Further tests with the model showed that, on average, the line was still only running at 78% of its required rate. Analysis of the detailed model performance statistics showed that the bottleneck was in the junction of the washer and load conveyor. The simple fix for the part collision had caused a part flow bottleneck.

The bottleneck resulted from part spreading, which occurred through the junction effectively reducing the flow of parts. The first part of a pair fell down the chute and landed at the bottom. When the second part reached the end of the washer conveyor it was turned off, since the first part did not have time to clear the landing zone. Once it did, the washer conveyor was restarted. The part on the washer conveyor moved for about half a part diameter before gravity took over, and it fell down the chute, taking another two seconds. Meanwhile, the first part was traveling down the load conveyor. By the time the second part landed on the load conveyor, what started as a 1 cm part separation became a 6 cm separation on the load conveyor, as seen in Figure 9. This spreading of parts, from repeatedly shutting down the washer conveyor, caused the reduced flow of parts through the junction.

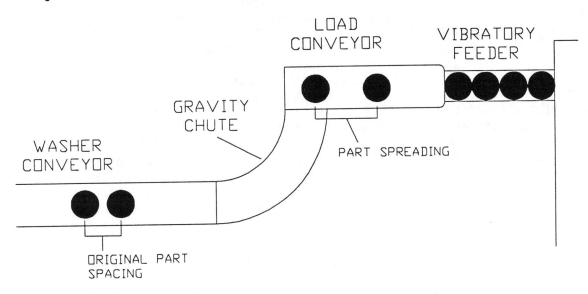

Figure 9. Part spreading on load conveyor.

This project illustrates the difficulties which can occur in multivendor lines. One vendor, designing a dual gripper for robotic loading, had invalidated the other vendor's control strategy, which assumed parts would be loaded more evenly spaced. These problems are difficult to detect prior to integration without a tool like XFaST (TM) to model the integrated system.

Initially, it was suggested that a stop be placed at the bottom of the chute to hold the second part nearer the landing zone. When the first part cleared, the second part could slide right in behind it. This would work, but the design team was concerned about adding another mechanism which could serve as a potential jam point for parts in the system. Alternative strategies were investigated for preventing part spreading without the addition of a stop. After several iterations, a control strategy was perfected solving the problem without additional sensors or mechanisms in the system. This ability to quickly prototype and test different control strategies by observing actual behavior allowed more optimal designs to be developed.

Since the washer and load conveyors were both instances of the same class of components (the SISO conveyor), other SISO junctions were checked for the problem. As expected, other junctions had similar problems with varying affects on the part flow. By changing the SISO class to incorporate the new strategy, these problems were solved as well. Furthermore, future applications using the SISO component will benefit from the more refined strategy by not having to deal with this problem again.

The washer and load conveyors were nonaccumulating conveyors. The vibratory feeder and accumulating conveyor were accumulating SISO conveyors. A different strategy was needed for these conveyors. Rather than just using a sensor to determine when the feeder was full, a part count was maintained as well. This eliminated the part spreading at the feeder input. It also gave the designers an opportunity to resolve a potential maintenance problem in the system. The original feeder control caused the load and washer conveyors to cycle on and off for every part. This placed additional wear on the conveyor motors, leading to frequent failures. With the part counter, it was possible to add some hysteresis in the conveyor cycling. Not turning the load conveyor on until several parts emptied out of the vibratory feeder, reduced conveyor on-off cycling by a factor of 12.

Table 1 summarizes the improvements made in the performance of different parts of the system. The final model showed that together, these improvements allowed the line to run 25% faster than required. It is important to note that the original model done, with a traditional simulation tool, did not predict any of these problems. Traditional models do not allow the detailed analysis of the actual control strategies used. Typically, they make simplifying assumptions about the behavior of conveyors and other material transport systems. XFaST (TM) models allow these problems to be detected in the design phase where the cost of correcting them, particularly capital equipment changes, is much less.

Table 1. Gear Weld Cell Throughput Improvements

Maximum Isolated Throughput)				
Device	Required (ppm)	Original (ppm)	Final (ppm)	Improvement
Robot	6.3	5.0	7.5	50%
Destacker	7.0	6.5	7.4	14%
Load	6.3	3.9	8.9	128%
Feeder	6.3	3.9	8.9	128%
Elevator	6.3	5.4	8.9	65%

Once the robot strategies were finalized, final production code could be produced from the detailed Grafcet designs. A mapping from Grafcet to Karel (the GMF robot language) was designed which allows this translation to occur directly. Although this is a manual process, it could be easily automated. Within a week, 1400 lines of fully debugged robot code (not including comments) were installed in the robot. The detailed tested designs and direct translation allowed for implementation to proceed more quickly than in traditional design cycles.

Modeling the operator determined whether a single operator could service the line. As it turned out, a single operator could not keep the line running 100% of the time. However, because the line is faster than required, production demands could still be satisfied with a single operator. The "control" design of the operator was done in Grafcet. When completed, the Grafcet diagram clearly depicted all the tasks the operator had to perform: the sequence of steps in a task, decision making, and strategies the operators must employ. This provides information allowing operators and plant personnel the opportunity to participate in improving the human factors of the line. Distances traveled, amount of weight lifted, operator consoles, and safety factors, can all be analyzed with these models to design a safer, better working environment.

CONCLUSION

These examples show the usefulness of XSpec (TM) methods and XFaST (TM) tools in designing manufacturing systems. Identifying errors in design and performance early in the life cycle allows quick changes. Different design trade-offs can be made more easily at this phase, since there has been no capital expenditure. Once a design has been tested with the tools, a rapid path exists for implementing the designs in the target hardware. In both of these applications, significant performance improvements were made possible through design changes. The cost savings in terms of increased production capacity, or reduced integration delays can run into millions of dollars. The XSpec (TM) method and the XFaST (TM) tools provide a solid basis for building the techniques needed to support the design of modern manufacturing systems.

Note:

ROBCAD is a trademark of Technomatix N.V.
FLEXIS (TM) is a trademark of SAVOIR.
SIMAN (TM) and Cinema are registered trademarks of Systems Modeling Corporation.
XSpec and XFaST are trademarks of the Industrial Technology Institute.

References

1. Ettlie, J.E., "Manufacturing Software Maintenance", Manufacturing Review, Vol. 2, No. 2, June 1989, pp. 129-133.

2. Alting, L., "Integration of Engineering Functions Disciplines in CIM", Annals of the CIRP, Vol. 35, No. 1, 1986, pp. 317-20.

3. Lieblein, E., "The Department of Defense Software Initiative - A Status Report", Communications of the ACM, Vol. 29, No. 8, August 1986, pp. 734-744.

4. Judd, R.P., R. VanderBok, M.E. Brown and J.A. Sauter, "Manufacturing System Design Methodology: Execute the Specification", Proceedings of CIMCON '90, May 1990, pp. 133-152, NIST Special Publication 785.

5. Sauter, J.A., R.P. Judd, R.S. VanderBok, and M.E. Brown, "XFaST - Integrated Tools for the Design and Rapid Prototyping of Manufacturing Systems", Proceedings of the International Workshop on Rapid Prototyping, IEEE Computer Society, Research Triangle Park, June 1990.

6. Parunak, H.V.D. and R. Judd, "LLAMA: A Layered Logical Architecture for Material Administration", Presented at the NATO Conference, June 1988.

Part 3

CIM Implementation Results

Ingersoll Milling
Machine Case History

by George J. Hess
Ingersoll Milling Machine Company

INTRODUCTION

The motivation behind our company adopting CIM technology was simple: we wanted to gain control of our business and improve our competitive position.

Winning the CASA/SME LEAD Award helped publicize the good work we were doing, and created a favorable image for us among our customers. We had over 1,000 people attend our free monthly tutorials, in which other companies, including most of our competitors, were allowed to hear a carefully orchestrated presentation describing the design, implementation, and operation of this system.

APPLICATION

This application is a computer-integrated enterprise for a medium-sized world-class manufacturing company. In defining computer-integrated enterprise, we include all of the following functions: the list of proposals, comprehensive customer master file, receipt of order, engineering design, release to manufacturing, procurement of materials, finish manufacturing, assembly, test, field installation, cost tracking, purchasing, inventory, accounts receivable, CAD, CAM, and comprehensive management reporting. The central database is said to be the largest in the world, performing nine million searches to gather information and serve each of these functions.

The technology includes the latest in mainframe and PC hardware, and solid-state disks. It provides video display units at nearly every work station that requires either graphics or alphanumeric data. It also uses the latest in software from the most sophisticated operating system on the computer to a full function database management system. An on-line, ad hoc inquiry capability to the data provides thousands of application programs, including sophisticated critical path method scheduling and remote communications capabilities.

WHAT COULD HAVE HAPPENED WITHOUT THE TECHNOLOGY?

From the time we started this project in the mid-1970s, our company has grown 10-fold. During this period, over half the machine tool companies in the United States have gone out of business. We have just completed building two of the largest machining centers in the world in Rockford, Illinois, and have installed them in Japan and Argentina. We needed the abilities of the computer to multiply our knowledge and efforts through this explosive growth not only in volume, but in technology. We could not have hired enough competent people to do all of this manually.

WHAT ARE THE BENEFITS OF THIS APPLICATION?

The first benefit is that everyone is able to work at a higher level of efficiency. They are able to work with dependable, high-integrity information. The old problem of conflicting and redundant information from various duplicate files has been eliminated. We were also able to do things we never could before, such as stress analysis calculations of machinery parts we designed to the optimum performance level. We no longer have to put in the safety factors that were necessary before.

When we installed the first computer graphics system, we built an economic simulation model of the company and predicted that we would start saving a million dollars a year, five years after its introduction. An audit after five years proved we were exceeding those initial goals. Other benefits, such as being able to train an N/C programmer in one month instead of 10, and being able to use the same geometric model in manufacturing that was constructed in engineering, have been tremendous benefits to our organization.

WHO ELSE CAN BENEFIT FROM SIMILAR EFFORTS?

This type of system can benefit all manufacturing businesses. Our system serves the engineering function very effectively with graphic design work stations. The results of this work is stored in the database, and called up by the manufacturing function to perform their numerical control programming, routing, and scheduling functions. These groups became the primary beneficiaries of these efforts. However, our sales organizations, and our customers also are served more efficiently by the accuracy and clarity of this information.

WHAT WERE OUR ORIGINAL GOALS?

Throughout this process, our original goals never changed. From the start, we wanted to produce the best machine tools in the world, to be the low-cost producer of those machine tools, and deliver them on time. Our goals are not related to specific systems, technology, or computers. We take advantage of technology where ever possible to meet our basic business goals.

HOW CLOSE HAVE WE COME TO REACHING THEM?

The best indication of our success is the acceptance of our products by our customers. Considering the tremendous growth of the company, it appears we are coming close to reaching our goals. However, this is a moving target. We have good competitors. They are constantly trying to do the same thing we are, so you never "reach" your goal. You must continually strive to reach the goals you set, and as you approach them, push them forward.

WHAT WERE THE NEGATIVE IMPACTS OF THIS EFFORT?

There were really no negative impacts, other than the cost. There were out of pocket costs for hardware, software, and management time. We have also had heavy management involvement. Training costs for people in all levels of the organization also were incurred. Still, when you consider our alternatives, this was one of the best investments we could have made.

OVERCOMING MANAGERIAL AND SOCIAL OBSTACLES

The social obstacles we had to overcome were the same encountered by everyone when change is introduced. We were asking each department to give up their proprietary information, and put it in a corporate database; we were asking each person to put away their pencil and papers and work at a keyboard and cathode ray tube; we were asking engineering designers to give up their drawing boards and work with light pens on the face of graphics work stations, and we were asking our manufacturing organization to put their highest skilled technicians in the office, preparing numerical control computer programs, rather than operate our multimillion dollar machines. We faced the same problems everyone does in this area, and probably did no better than anyone else in overcoming them. We were dealing with problems involving "human nature." Nevertheless, we managed to overcome them.

Managerial obstacles were few, largely because of the leadership of the top executives in our company. They invested the time and effort to learn the true capabilities and limitations of these kinds of systems. They were willing to accept failures, and learn from it without hesitation and with full support for the project from day one.

Even when business was bad, we mortgaged some of our major assets to buy new computers and new graphic work stations. This is an example of outstanding resolve.

CONCLUSION

Our application is no longer unique. Surely, every manufacturing business that has survived the onslaught of international competition must, by now, have been seriously working on a computer-integrated enterprise. However, we feel that we were original in our field. We know of no other manufacturer that was pushing as aggressively, or as early as we were with the clear vision of the future that we had set down in our master plan. Our efforts were rewarded when, in 1982, our company became a recipient of the CASA/SME LEAD Award.

Deere & Company's CIM Evolution

By William G. Rankin
Deere & Company

INTRODUCTION

Deere & Company is the world's largest producer of agricultural equipment. It is recognized for its quality products and innovation in manufacturing. Most noteworthy, was its selection by the Computer and Automated Systems Association of the Society of Manufacturing Engineers, as the first annual LEAD Award recipient for leadership and excellence in the application and development of computer-integrated manufacturing.

This paper reviews the automated tractor assembly facility project for which Deere won the 1981 LEAD Award. It also examines the lessons learned from the effort, and the CIM directions Deere has taken since.

THE JOHN DEERE TRACTOR WORKS

During the early 1970s, Deere initiated a major expansion and modernization program for its manufacturing complex in Waterloo, Iowa. One of the most significant projects in this program was the design, development, and implementation of a new tractor assembly plant.

The project had three primary objectives: to provide superior products that consistently deliver on the promise of their advanced design - a growing challenge with today's larger more complex machines; to assembly processes and support functions, and to provide employees with a superior work environment.

A project management team was formed in early 1976 with representatives from the existing factory and corporate headquarters. This multidisciplined mix of practical experience and state-of-the-art knowledge resulted in sound engineering analysis and the adoption of technology that works. Numerous alternatives of varying degrees of automation were evaluated. In most cases, the highly computer-integrated alternative provided the best return on investment.

The resulting 2.1 million square foot complex, designated the John Deere Tractor Works, became fully operational in the spring of 1981. It is made up of seven buildings, as shown in Figure 1, and includes an office building, an energy center, a receiving facility, a Sound-Gard Body

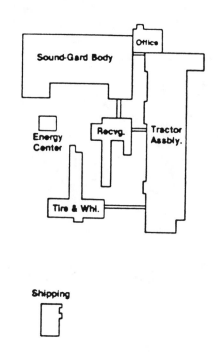

Figure 1. Deere and Company building layout.

fabrication building, a wheel manufacturing building, the tractor assembly building, and a shipping facility. A network of minicomputers and a host mainframe coordinate and control the flow of material through the operation. They request parts from inventory, deliver them, notify other computers that they are coming, replenish inventory as required, issue build orders for subassemblies and components, and integrate the assembly line activities so the right part arrives at the right place at the right time.

All operations are driven by the line-up or sequence of tractors on the assembly line. Tractor build orders are run overnight, with the mainframe host computer communicating requirements to the minicomputers.

MINICOMPUTER APPLICATIONS

There are nine Digital Equipment Corporation (DEC) minicomputers controlling various aspects of operations at the tractor works. These are diagrammed in Figure 2, along with their positions in the buildings where they are located. The following describes each with emphasis given to their process control functions.

The first minicomputer is for the receiving facility. A minicomputer controls the conveyors that route incoming loads to a high-rise storage facility containing 16,000 locations. The minicomputer directs the stacking cranes in placing loads in the storage facility. As material is needed in production, the computer directs its retrieval from the high-rise.

The second minicomputer is for the Sound-Gard Body building. Here, workers fabricate, paint and assemble the self-contained operator stations which house all controls, the seat and rollover protection for tractors. This second computer is referred to as the Work-in-Process/Welded Painted Body

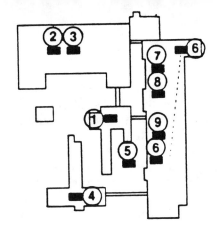

1. Receiving/Stores
2. WIP/WPB Stores
3. Paint & Conveyor
4. Tire & Wheel Stores
5. Interbuilding
   Delivery Conveyors
6. Chassis Assembly
   Storage/Finish
   Trim Storage
7. Tractor Assembly
   Conveyors
8. Finish Assembly
   Storage
9. Repair Tractor
   Tracking System

Figure 2. Diagram showing the positions of minicomputers controlling the tractor works.

Storage System. This is a high-rise storage system similar to the receiving facility, except it is used to store work-in-process inventory. There are actually two separate storage facilities. In addition to the main storage high-rise containing 7,000 locations for work-in-process inventory, there is a second storage high-rise which has the capacity to store and retrieve 132 welded and painted Sound-Gard bodies.

The Sound-Gard Body building also contains a third minicomputer, referred to as the Paint and Conveyor System. This is a conveyor system under computer control which routes parts and subassemblies through manufacturing and paint areas. The minicomputer directs the routing of parts to the correct color and type of painting operation and controls the speed of the conveyor. This minicomputer also monitors approximately 500 key data items in the paint system to ensure a quality painting operation.

The fourth minicomputer is dedicated to the tire and wheel building. Wheel manufacturing is like a minifactory all by itself. Wheel parts are received, stored, retrieved, machined, and assembled with tires to support specific tractor assembly requirements. This fourth minicomputer controls the high-rise system. It is similar to the other high-rises except it is dedicated to tires and wheels. This high-rise contains approximately 12,000 locations.

The fifth minicomputer controls the interbuilding delivery conveyors. The buildings are connected by corridors which house a conveyor system under the direction of a minicomputer. The minicomputer controls the movement of goods on this conveyor between various load and unload points in the system.

The sixth minicomputer is the first of four dedicated to the tractor assembly building. This computer controls two high-rise storage systems. The first system is for chassis assembly components stored in the northeast corner of the building. This high-rise contains 1,240 locations for the storage of assembled engines and transmissions with axles. The same minicomputer also controls a second system. This is a high-rise storage facility located near the end of the assembly lines containing approximately 400 locations. This high-rise stores finished trim appearance parts.

The seventh minicomputer controls the conveyors in the tractor assembly area. This minicomputer moves the tractor assemblies through the assembly process. This is one of the most complex systems based on communication requirements with other computers.

The eighth minicomputer controls the third high-rise storage system located in the assembly building. This is the Finish Assembly Storage System. It contains approximately 1,000 locations for fully assembled Sound-Gard bodies and other parts used on the chassis assembly lines.

The ninth minicomputer also supports the tractor assembly building. It is a minicomputer which controls the flow of finished tractors through the final adjustment and repair area located in the south half of the assembly building. The purpose of this computer is to keep an accurate record of all tractors in this final adjustment area prior to finish trim. This computer controls these tractors in a first-in/first-out sequence based on their scheduled build dates.

These nine minicomputers operate 10 systems to control the manufacturing processes and movement of materials throughout the facility. Minicomputers play a major role in nearly all manufacturing and assembly operations at the Tractor Works. Using computers for process control is nothing new. What makes the Tractor Works unique is that the minicomputers in use are also data processing and business management computers.

COMPUTER COMMUNICATIONS

The 10 systems running on the nine minicomputers are shown in Figure 3. Also shown is the IBM host computer located in Deere's remote computer center. The 10 systems are identified by abbreviations. For example, REC stands for the Receiving High-Rise Storage System. The computer center is identified by the term HOST.

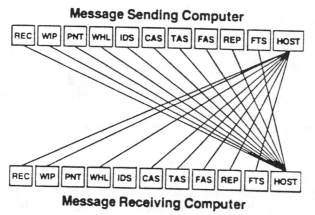

Figure 3. Ten systems controlling manufacturing and material movement throughout the facility.

The 11 blocks across the top of the figure are designated as message sending computers. These same 11 blocks on the bottom of the figure are designated as message receiving computers. Lines drawn between systems indicate messages are transmitted from the message sending computer to the message receiving computer. As depicted, all of the minicomputers send messages to the IBM Host. In most cases, these messages are to update master inventory records maintained on the host computer. The host computer also sends messages to minicomputers. Only the paint computer identified as PNT and the Interbuilding Delivery System or IDS computer do not receive messages from the host. The messages sent to minicomputers are primarily in two categories. The first is from computer programs run overnight telling the various high-rise systems what material to withdraw from storage the following day. The second category of messages represent build orders to various computers for wheels, axles, and completed tractors. What is really happening, is that the host computer is running the business, just as it does in traditional manufacturing environments. The difference is it is telling minicomputers instead of people what output is expected from them each day.

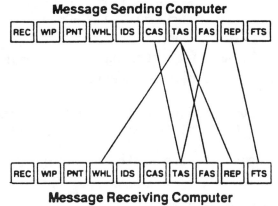

Figure 4. The minicomputers can communicate with each other.

Figure 4 shows another unique aspect of the Tractor Works operation. In addition to communicating with the IBM host computer, the minicomputers also communicate with one another. Messages transmitted are basically of the category, "I need a part. Do you have it?" This question is analyzed and answered, and the part is sent on its way if it is available. These messages are transmitted to make certain a tractor does not get started down the assembly line unless all major components are available for assembly. None of the first five computers shown initiate a message to another minicomputer. All possible messages cannot be described here. However, the following is provided as an example:

The CAS (Chassis Assembly Storage System) withdraws engine and transmission assemblies from storage. It sends a message to TAS (the Tractor Assembly System) telling TAS an engine and transmission are coming. TAS then sends a number of messages. One is to FAS, our Final Assembly High-Rise. FAS then sends the appropriate message to TAS indicating whether or not the Sound-Gard Body is coming. The TAS or Tractor Assembly Computer then decides if it will proceed to build the tractor.

Each of the minicomputer systems function independent of the other. However, if one system goes down, the next one is affected. This is avoided for short durations by maintaining inventory supply areas between systems. This situation is really no different from conventional manufacturing processes except that minicomputers are involved.

Figure 5 provides a recap of a few of the sequential events that take place under computer control when a tractor order is received at the Tractor Works. These events are briefly described with the numbers corresponding to the sequential numbers in the figure.

1. In receiving, purchased Sound-Gard Body material is withdrawn from storage and sent to the Sound-Gard Body building.

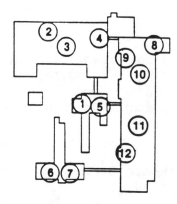

Figure 5.   Some sequential events occurring under computer control when an order is received.

2. In the Sound-Gard Body building, materials are withdrawn from work-in-process storage and sent to manufacturing areas.

3. Sound-Gard Body weldments are washed and painted under computer control.

4. The assembled Sound-Gard Body is sent to the tractor assembly storage system.

5. Back in receiving, other purchased material is withdrawn from storage and sent to the tractor assembly building.

6. In the wheel building, tire and wheel parts are withdrawn from stores for assembly.

7. Completed tire and wheel assemblies are then sent to the tractor assembly building.

8. At the north end of the assembly building, engines and axle and transmission assemblies are withdrawn from storage and sent to the tractor assembly lines.

9. The tractor assembly moves down the line, its status being constantly reported to the minicomputer and to the host computer.

10. The tractor assembly moves down the line, its status being constantly reported to the minicomputer and to the host computer.

11. The tractor moves through the tractor adjustment area again being traced by a minicomputer.

12. The tractor is processed through the finish trim operations and moves to shipping. It is at this point that the tractor is finally released from minicomputer control. Of course, the host computer still knows the location of the tractor.

As this sequence of events indicates, there is very little that takes place at the Tractor Works in terms of material storage and movement that is not done under minicomputer control.

PROJECT EXPERIENCES

When the John Deere Tractor Works facility project was initiated in early 1976, there was no intention of creating a highly automated, computer-integrated manufacturing plant. Personnel within Deere had little experience with computer controlled manufacturing, and there were few if any computer-integrated facilities in existence to learn from. What Deere did have experience in was project management. The project was well organized and broken down into manageable pieces corresponding to the functions of the resulting buildings. Visits by project teams to various manufacturing facilities clearly demonstrated the effective use of minicomputers in high-rise storage and conveyor applications. Because there was no master plan for integrating computerized material handling functions, minicomputer systems were considered as separate entities within individual construction contracts.

During 1977, the teams recognized the full potential of computerization within their project. They began thinking of computerized material handling systems as part of an integrated whole. However, by this time, one system was almost ready for delivery and other contracts had been awarded. It became obvious that the various material handling systems should interface with each other and with the host system. By the time this requirement was recognized, the teams were dealing with a variety of different vendors who had little or no experience integrating with the systems of other vendors. Deere's systems personnel concentrated on developing the required new host computer systems and specific interfaces to the mini-systems. Additional work went into defining the specific functions of the mini-systems and mini to mini interfaces. Of course, the major effort of the project team focused on engineering, the manufacturing and assembly processes, and the facilities. The receiving building was the first to become operational with startup in November 1977.

During 1978, functional definition and clarification of all systems continued. Standards were issued to vendors concerning specific pieces of DEC hardware and software. A central computer room concept was adopted and numerous variations of computer backup were investigated. All remaining systems were installed during 1979 and 1980. The facility became fully operational in May 1981.

LESSONS LEARNED

The creation of the John Deere Tractor Works was a pioneering effort and a major achievement. Much was accomplished and learned. The objectives established in the original planning are being achieved. Superior products are being built with the precision assembly processes that computer control can provide. Tractors are being assembled more efficiently than ever. Overall tractor work-in-process inventories have been reduced by 50%, tires and wheels by two-thirds and raw steel to one-sixth of the original level. Lead times are shorter and production is more predictable because there are fewer surprises. Employees have been provided with a clean, comfortable, pleasant working environment. Undesirable jobs, such as painting, have been automated.

In addition to its planned objectives, the project also provided Deere with the opportunity to experience and learn about the implementation of CIM.

Deere learned first hand that there is no such thing as a turnkey system. If a vendor's system is to be integrated into an existing Deere application or computer network, it will require modification.

Another lesson learned was the importance of establishing detailed standards, or at least reducing the variability among vendors in such areas as system architecture, application techniques, process logic controllers, cabling, and wiring schemes. Line drivers, diagnostic tools, models of terminals, pin configurations, system generation options, documentation, and user manuals are other areas requiring less variation.

An appreciation was also gained for the critical importance of having adequate in-house systems expertise. Programmers and analysts needed to be trained from the ground up for the DEC systems. Detailed software systems knowledge was required for application development and maintenance. Lack of this knowledge made it difficult to monitor and participate in system design with vendors. Inadequate training, standardization, and documentation made the transition of support from vendor to in-house personnel more difficult.

Deere also gained an understanding of the importance of planning for system maintenance and future enhancements. During implementation and interfacing for integration, changes were made causing the plant to operate slightly different than planned, and systems had to be modified accordingly. In some cases, requirements for improved system performance made it necessary to upgrade to more powerful computers.

CIM DIRECTIONS

The John Deere Tractor Works project provided the company with a unique opportunity - the creation of a new plant incorporating state-of-the-art computer technology and an integrated manufacturing strategy.

Opportunities for new plant startups are rare, and the majority of manufacturers today face the challenge of achieving CIM within existing facilities. John Deere is no different. Most of its CIM thrusts are aimed at this objective. Analytical tools and system components of CIM are being developed which are modular, hardware independent, and functionally oriented. These tools and systems are being linked and selectively applied at nearly all Deere factories.

As a fundamental step toward achieving CIM, Deere has been reorganizing many of its large, complex, manufacturing operations into focused, simplified cells. These manufacturing cells are the basic building blocks of Just-In-Time manufacturing and provide manageable opportunities for computer integration. The John Deere Group Technology System is the key to this reorganization effort. It allows grouping various machines and processes into cells in which families of parts are manufactured complete. This state-of-the-art system, which Deere is now marketing, provides a comprehensive part database and a powerful set of software tools to extract and analyze part features and form logical part families.

In addition to group technology, other systems basic to CIM are being developed and integrated. Group technology's feature based representation of parts provides a bridge between CAD and CAM. It is fundamental to Computer-Aided Process Planning (CAPP). Designed as a knowledge-based system, CAPP allows the development and application of consistent and standard logic for processing a part. In addition to the part feature data in group technology, CAPP uses similar databases about machine tools and tooling in deriving manufacturing plans and costs.

To provide integration with Geometric Modeling Systems (GMS), the group technology database is being extended to a complete feature-based part description in a neutral format. This provides a more complete, higher level description of part features, such as holes, notches, slots, etc., than the pure geometry generated by GMS. The resulting data can be used more directly by down stream applications, such as CAPP. Efforts are underway to automate the creation of this data as part of the geometric design process. The group technology feature database is being used by designers to retrieve existing designs, avoiding the designing of new parts, and standardizing on selected features. The concept of designing parts by inserting parametrically defined features also is being investigated.

The integration of engineering data with production data, and providing it in an accurate and timely manner also is critical. A set of databases and systems referred to as COMMON have been developed and implemented to achieve this objective. The COMMON Specification database contains product structures; COMMON manufacturing engineering contains most of the information generated by process planning and other manufacturing engineering functions and makes this data available to other areas and the shop floor on a real-time bases; COMMON MRP and COMMON Master Schedule are used to determining production schedules and material requirements; and the COMMON Work Force - Machine Load system aids in developing machine and manning requirements. Other COMMON systems and databases in use or under development address the managing and tracking of interfactory and supplier part orders and deliveries and purchasing functions. The COMMON systems and databases provide the medium for integrating much of the information required for the factory of the future. Many other tools, such as group technology and CAPP, are interfaced to COMMON to retrieve and deposit information and to more easily integrate applications. These systems are being migrated to relational databases.

Shop floor automation requires the distribution of data and systems and the establishment of a communications media and protocols capable of interfacing a variety of different devices. Today, Deere factories have stand-alone computer numerical control (CNC) machine tools. There are also robots used in a variety of applications, highly sophisticated flexible manufacturing systems (FMS), and numerous shop floor and cell control computers. These are used to control processes, support production control and attendance and production reporting functions. Nearly all new machines purchased are CNC, enabling them to be linked with planning and control systems via distributed numerical control (DNC). To facilitate the communications necessary to achieve CIM on the shop floor, some Deere units have installed local area networks (LANS). The most recent have been in the form of broadband cable backbones. This will enable a migration to a standard communication environment based on the General Motors Manufacturing Automation Protocol (MAP) specification. Deere actively supports MAP and is a founding member of the MAP Steering Committee. MAP pilots are in place to support CIM applications at two factories.

Use of other standards (de facto or emerging), is a key element in ensuring cost-effective integration of all of the various applications suitable to a particular installation. UNIX operating systems and SQL-compatible relational database management systems, for example, are de-facto standards which will maximize the portability and life spans of software as hardware continues to evolve.

Integrating the diversity of computerized tools, systems and databases through the functions of product design and drafting, production planning and production execution for a large existing manufacturing facility is a monumental undertaking. To make this task feasible, Deere has used pilot, or prototype projects to implement CIM within the cells discussed earlier. Each cell represents a vertical slice through a factory. The family of parts manufactured within the cell and the cell itself are treated as a minifactory. All of the functions from conceptual design through physical manufacture can be better understood and the tools and systems can be tailored and integrated to achieve CIM, or this factory within a factory. By building the tools and systems to be modular and portable, they can be tailored and used for other applications.

This approach of simplification before integration helps avoid misapplying computer technology to better manage unnecessary complexity. It allows CIM to be achieved in digestible bites. It has allowed Deere to effectively incorporate rapidly changing technology into its manufacturing operations and maintain its competitiveness and leadership position.

CIM Contributes to Profitability at Allen-Bradley

by John Rothwell
Allen-Bradley

INTRODUCTION

In 1988, the Allen-Bradley Company won the CASA/SME LEAD Award, one of the most prestigious awards a manufacturing operation can receive. Winning it helped promote Allen-Bradley's increasing presence in the global factory automation marketplace.

The implementation of CIM in our long-term strategic manufacturing plan was based on a decision to expand the company globally. Our objective was to produce a world-class product to compete in an extremely cost competitive market. Only through the use of CIM could we produce these products profitably. As we introduce other products into world markets, we are applying CIM technology wherever feasible. When we introduce a new product into manufacturing, the contribution it makes to our profitability is examined. A target cost is determined through market research which examines the competitive cost worldwide. Setting the target cost enables us to be competitive globally. In many cases, CIM has proven to be the only method of manufacturing that allows us to meet this target cost and remain competitive globally.

Using CIM to produce contactors and relays has improved our market share of IEC devices by 15% beyond market projections during the past five years. We believe we are the lowest cost producer of these products worldwide.

The CIM system enables an Allen-Bradley salesperson anywhere in the world at any time of day to enter a customer order and have it shipped from this facility within 24 hours. Today, the system produces 849 different contactors and relays in two frame sizes in a lot size as small as one, and at a production line rate of 600 units per hour with only a few attendants overseeing the process.

Since it opened in April 1985, over 25,000 customers, students, and professional groups have toured the 45,000 square foot facility. This has given Allen-Bradley an opportunity to show how existing facilities can be automated.

SYSTEM DESCRIPTION

Each of the 26 machines that assemble, test, identify, and package products is controlled by an Allen-Bradley PLC-2/30 programmable logic controller, which communicates with an A-B PLC-3 cell controller through two A-B Data Highways. The PLC-3 controller is linked to an A-B Pyramid Integrator area controller through another A-B Data Highway. The Pyramid Integrator is interfaced to our IBM 3090 mainframe computer with an Ethernet link using an Interlink gateway. Our system integrates every facet of our organization from order entry to the shop floor, including sales, marketing, engineering, operations, quality, finance, and management. The design philosophy was to gather all the data and place it in a database, so that marketing, sales, engineering, quality assurance, and other functions could use the information. This information is used for sales order entry, market forecasting, production and inventory control, cost accounting, statistical quality control, shop floor monitoring, maintenance diagnostics, and management reporting. Although our CIM system is complete, we plan to develop future applications as computer technology continues to grow.

Since our facility was designed in 1983 and 1984 when CIM technology was still in its infancy, we were world-class innovators in every facet of the project. We began with a computer-driven project management system. This allowed us to track and delegate project responsibility. By designing the product and the process simultaneously using CAD/CAM, we "designed for automation." The system was designed with the manufacturing philosophy of Just-in-Time, batch-of-one, and 24-hour build to customer order and shipment of production. Prior to this, our production of a similar product had been manual assembly using weekly production schedules built for stock. Innovative programmable technology, such as laser-etching and ink-jet printing enabled us to meet our objectives. A-B's Productivity Pyramid concept of CIM is proven through the integration of four levels of computers from our IBM mainframe down to the factory floor and back to form a closed-loop information feedback system.

RESULTS

Applying CIM technology had only positive results for the Allen-Bradley Company. Our plant has been recognized as one of the finest examples of leadership and excellence in the application and development of CIM worldwide. We have once again made manufacturing a competitive weapon.

WHAT WE WOULD DO DIFFERENTLY

Many of our customers who visit our facility ask, "If you had to do it over again, what would you do differently?" We tell them that we would design the software to do only the essential requirements of the system. If the software is under utilized or found to be superfluous, time and money are lost. Additional software can be written to accommodate needs as they arise. Memory is inexpensive, software is not.

In the beginning, we did not have a computerized maintenance system running in our contactor facility. We have it today and found it to be a necessary ingredient for any automation system. Performing preventative maintenance based on usage rather than time is a real benefit to uptime, throughput, and

longevity of the machinery. The common database we use allows us to add
new maintenance requirements when something new is learned about the system.
Selecting the most qualified maintenance personnel to work on the project
during implementation promotes ownership and understanding. Once the system
is understood, listing and maintaining essential spare parts at start up
just makes good sense.

Ensuring 100% good incoming part quality also adds to the success of
automation. It took several years working with vendors and suppliers of
piece parts to ensure SPC principles were being followed. SPC is a must,
and only adds to better outgoing quality. The old saying "garbage
in/garbage out" certainly applies to any CIM system. Demanding PPM reject
ratios of 50 or better from all vendors only enhances our zero reject
philosophy.

WHY DID THIS PROJECT SUCCEED?

Success of our project can be attributed to many factors. One is the
definition of the functional specification. The project began with a
multidisciplined 27 member task force whose objective was to write the
system's functional specification. Leading the project was the vice
president of operations, who requested the help of directors and managers
from marketing, sales, production and inventory control, facilities
engineering, quality, development, automation design, purchasing,
manufacturing, manufacturing engineering, finance, and other areas.

The automated equipment was designed by A-B's mechanical and electrical
engineers, using simultaneous engineering. Development engineers
concentrated on product design, while manufacturing, process, and tool
design engineers configured the processes to be used.

Another key success factor in our project was its champion, Larry Yost, vice
president of operations. The decision to implement CIM came from our
business strategic plan. This was a new product entering a new market, and
to be a survivor in the IEC marketplace, we had to be the lowest cost
producer of our product. The risk was great. The project took two years to
design and build, at a cost of over $15 million for equipment, computer
hardware, and software. Larry recognized the need and risk, and took a
leadership role in communicating our position to salaried and hourly
employees. It was through this total top-down commitment that involvement
in CIM implementation became a way of life with our employees. The
leadership of the implementation task force played a critical role in
bringing the project in on time, and within budget. Members of the task
force included representatives from production management, systems design,
marketing, production and inventory control, facilities engineering,
quality, automation design, and manufacturing engineering. Middle
management was asked to provide the best people to ensure the success of the
project.

ADVICE FOR OTHERS

Some advice we share with others applying CIM is as follows:

   1. Know your company and its future business strategic plan.

2. Select the best people for the project.
3. Define functional specification with input from all disciplines.
4. Develop the product and the process simultaneously.
5. Delegate responsibility and authority, and hold people accountable for their actions.
6. Pay attention to detail, and be a perfectionist.

Allen-Bradley's CIM facility was the largest single manufacturing project undertaken by our firm to date. The original plan was to design and build the system in three years. This was the first time we had designed the product and the process simultaneously.

Development of software required over 12 staff-years of programming time. The greatest innovation was the design of "Software Backplane," structured to be reusable, transportable, and modular. It is now marketed for CIM applications worldwide. The equipment consists of 26 fully automated assembly machines. These machines which assemble, test, and package were principally designed and built at Allen-Bradley and employ laser-bar code readers, programmable ink-jet printing, laser-gaging, vision inspection, laser-etching, and numerous types of sensors. Over 150 highly skilled Allen-Bradley craftsmen helped to build and install the system. The LEAD Award was more than just an award; it was a tribute to all those employees within the enterprise who helped "make it happen." It gave the team the greatest recognition imaginable.

Part Four

Japanese CIM Installation

The Total CIM System for Semiconductor Plants

by Yasuo Mizokami
OKI Electric Industry Co., Ltd.

INTRODUCTION

Conventional semiconductor plants have built up their capability using independent computer systems in production control, quality control and automation. The quality and productivity of LSI plants has been stepped up year by year, in spite of increasing numbers of devices with more complicated processes. In the most advanced plants, these independent computer systems are integrated as a total system for covering total plant operation. This has been established in our latest plant for memory device production. Such a system consists of six sub-systems.

PRODUCTION CONTROL SYSTEM

Referring to market situation, sales department order status, and plant congestion, this system creates a schedule for suitable production. It instructs the automated transportation system to distribute individual wafer lots to the appropriate process step in the plant. This is based on information like products specifications, line monitor, equipment status, and production schedules.

These independent systems are installed at each plant. They are hooked up to the computer system situated at the headquarters to monitor and control the whole production status.

TECHNICAL INFORMATION SYSTEM

The engineering group can give precise instructions to the related plants for product information, such as product specification, material, mask information, etc. It can also deliver engineering standards like process and test specifications through the system.

PROCESS QUALITY CONTROL SYSTEM

This system provides each process step with instructions to produce each wafer lot based on information from every process stage, the condition of

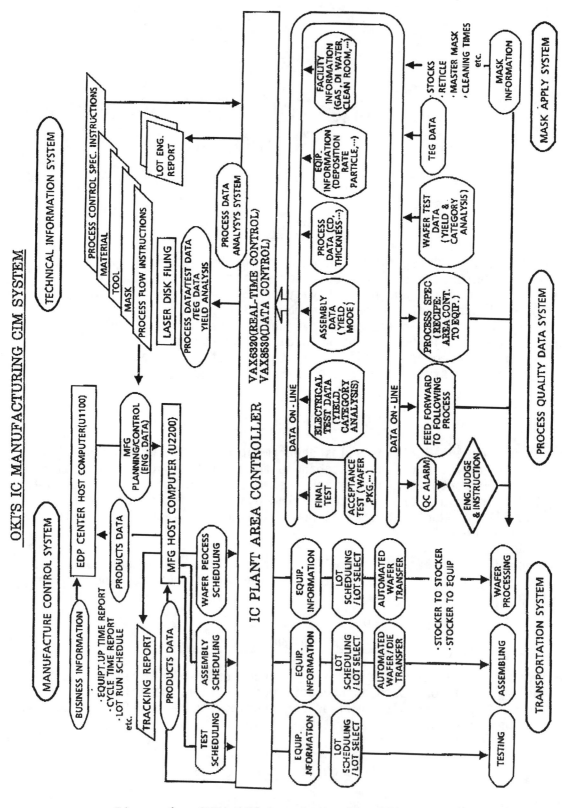

Figure 1. OKI's IC manufacturing CIM system

equipment and facilities, as well as process flow and specification. This system can handle SQC at every major process step as well.

PROCESS DATA ANALYSIS SYSTEM

When product failure or irregular mode occurs, or when yield analysis will take place, this system will help analyze the cause, based on the process date.

MASK FABRICATION LINE OPERATION SYSTEM

CAD systems and Electron Beam Exposure systems are connected by this system. The system is capable of automatic JDF, quality and production control in the mask fabrication line, and mask maintenance line.

AUTOMATIC TRANSPORTATION SYSTEM

Our wafer process plant is split into blocks containing photolithography, etching, oxidation and diffusion, ion implantation, CVD, metalization and wet cleaning, etc. Each block has its own automated stocker for transporting wafers by linear motor vehicles situated between blocks. Wafers are delivered from the automated stocker to production equipment through the sublinear motor vehicle system running on the ceiling. This is the Wafer Tracking Control System. It controls the Automated Transportation System based on production control, and equipment status information.

REVIEW AND DISCUSSION

Employment of this system is leading to prevention of operational errors, reduced manpower, and improved quality control. In addition, thanks to installation of the ultra-clean process, we have realized a high yield VLSI plant with less operators. We still have, however, many issues to improve upon.

AUTOMATION

Due to the low operation rate of the most advanced equipment for semiconductor production, the automated line frequently cannot give full scope to its function. It is important to improve such equipment through TPM activities, as well as having vendors improving activities. Since development and improvement of computer systems requires large investment and numerous engineers, it may sometimes be hard to achieve the expected economical effect.

QUALITY CONTROL

Under Japanese TQC activities, one of the most important issues is that operators have direct access to raw data. Supervisors often question the practice of keeping operators isolated from raw data because of the heavy use of computer systems in the plant.

STABILITY OF PROCESS

Often, new processes and devices are not completely debugged before being released from research and development labs to advanced LSI plants. Under these circumstances, it is important that engineers improve the margin and process window on their design and processes.

Part Five

The Product Data Exchange Standard

PDES:
The Enterprise Data Standard

by Robert Carringer
Institute of Business Technology

INTRODUCTION

The Product Data Exchange Standard (PDES) is a data description and format standard under development for the exchange and sharing of data needed to describe a product and its manufacturing processes. This paper provides a general introduction to the technology, development plans, organizations and benefits of PDES.

WHAT IS PDES?

PDES can be viewed from several different viewpoints and is actually a compilation of many different activities. PDES is a standards development process. It is a combination of different technologies, contributed by people from many companies and organizations.

As a standards development process, PDES is an activity with a goal of creating an international standard for the exchange of product model data. The resulting standard is also a process whereby knowledge is created, shared and documented. The standards development process is unusual because the standards activity is happening concurrently with the development of the technology. In most standardization processes, the technology is developed, stabilized and then standardized. The PDES program is doing these concurrently.

The technology going into the development of the PDES program can also be categorized in several different groups.

The first technology to be discussed is information modeling. Here, the purpose is to define and document the product data to be shared or exchanged between organizations and functions. Information modeling has become the modeling of knowledge, which includes:

* Knowledge about the enterprise,
* Knowledge about the products produced by the enterprise,
* Knowledge about the process used to produce those products.

Once the information has been modeled, there must be a way of storing and retrieving it. Information management technology is also an important part of the development of PDES. Information management is being stressed because of the different information used in the PDES program. Information management technology has to provide access to the PDES information, storage and retrieval of the PDES data, the ability to create new PDES data and modify it.

The third category of technology is downstream technology that uses PDES data. PDES is frequently tied to downstream technology to demonstrate its capabilities. For example, a computer-aided process planning system, an example of downstream technology, is often an information starved application needing a PDES database to become fully automated. From the history of the development of PDES, we find that downstream applications provide excellent sources for the requirements of a PDES database, or a PDES system. Those downstream applications like computer-aided process planning, group technology, classification and coding, machine parts programming, robotics programming, inspection system programming, all require PDES data as input, and can also store back data into the PDES database.

These are all groups of technologies which together become part of the PDES program, and upon which the PDES program relies, and through which the PDES program will be demonstrated to the manufacturing and engineering communities.

PDES is also an activity. The PDES program started in 1984 as a spin-off of the IGES activity. It has been administered by the National Institute for Standards and Technology (NIST). NIST administers the PDES program through its IGES/PDES organization. The people involved with PDES are from companies throughout the world. They volunteer their time and company resources to work on the development of the PDES specification and on the standardization of PDES.

The activity of PDES can also be traced back to technology development projects funded by the Air Force Manufacturing Technology Program. Research and development programs have been funded by the Air Force since the early 1980s working on the technologies underlying PDES and other activities related to PDES.

When asked, "what is PDES?" we get several answers: PDES is a standards development process; a goal of the PDES program is to create an international standard for the exchange and sharing of product model data; PDES is also a technology; PDES is a technology of information management, modeling, user technologies and implementation technologies related to the overall industry of CAD/CAM and CIM. And, finally, PDES is an activity contributed to by companies around the world, by U.S. government agencies, the Air Force, and other military branches of the Department of Defense.

These different viewpoints make possible this description and explanation of the PDES program, so that your company may become involved in the standards development process, understand the technologies related to the PDES program, and become involved in PDES, the activity.

The requirements for PDES and all its related activities are:

* Define the product and process of manufactured systems;
* Support exchange and sharing of product information with a minimum of human interpretation;

* Interrelate a broad range of product information to support
  applications found throughout the product life cycle;
* Define a single, logical representation of product information and
  application views (content and format).

The product and process information can be grouped into categories. The
information is modeled using graphical and computer language techniques.
The resultant topical data models contain information related to: geometry,
solids, tolerances, electrical functions, material, presentation,
architectural, topology, form features, layered electrical products, finite
element modeling, product structure, drafting, and ship structures.

Information modeling is based on a three-layered approach. The logical
layer is a definition of all the information outlined above. It is a single
data model or representation of the data. Also called a conceptual layer,
the logical layer is the central model from which the other two layers are
derived. The application layer is a subset of the logical layer defining
the information for a particular application (design, process planning,
inspection, etc.). The physical layer is the definition of the actual file
or database containing the PDES data. The PDES physical file format is an
example of the physical layer.

A HISTORICAL PERSPECTIVE ON PDES

PDES has as its objective to electronically define all the information
needed to design, manufacture, and support a product. We find some of the
early demonstrations of PDES technology relate to being able to define the
product's geometry in a computer system, demonstrate the automatic creation
of machining instructions, and machine the product that fits that initial
product definition geometry. That, as an objective, can be traced back to
several things in the past, going as far back as the development of
numerical control in the 1950s.

The original objective of the NC development program was to automatically
define a part so that control of part manufacturing moved off the shop
floor to an engineer in an office. The engineer could control how the
machine ran and the products the machine makes. That objective was
partially realized in the development of APT. We find in APT the ability to
define simple geometry and tool paths to manufacture specified geometry.

Defining products in APT was really limited to the simple geometry modeler
within an APT program. It was not sufficient to support other downstream
applications and other upstream applications to create the geometry in the
first place.

Through the 1960s, we have seen the implementation of graphics systems
within manufacturing organizations. These systems define and model
products, and have become computer-aided drafting systems. There is also
storage of information about a product. Throughout the 1960s and
1970s, the development of CAD/CAM evolved in such a manner. We are defining
products, we are moving from the ability to put out an automated blue print
to being able to put out geometry information that can drive downstream
applications like NC part programming. Once we obtain NC part programming,
we want to use it for things like coordinate measuring machine instruction,

and also upstream into the design process to be able to perform design analysis on this CAD/CAM data as well.

In the late 1970s addressing the need of exchanging CAD/CAM data between CAD/CAM systems, IGES was invented by Boeing, General Electric and the Air Force. The objective of IGES is, again, the ability to exchange a product definition between two different systems. And why do we want to do that? Because we want this information about the total life cycle of the product to become digital and we want to exchange digital information or electronic product description data instead of paper drawings.

From our experience with IGES in the early 1980s, the manufacturing engineering community saw there was a need for other information to be stored electronically. Information like the manufacturing intent of a product, or the manufacturing form features, as we later called them. In viewing information as it is released from engineering to manufacturing, we see there are several categories of information actually released from design or desired to be released from design, to support applications and manufacturing.

The Air Force funded a program, the Product Definition Data Interface, whose objective was to define information needed by manufacturing, created in engineering, to support manufacturing applications.

The PDDI program presented its early results to the IGES organization in 1984. The IGES organization formed the PDES initiation activity. That was the first time the IGES organization chartered any activity with relation to PDES and, in fact, was the early genesis of the PDES program.

Design automation, manufacturing automation, the implementation of computers to design and the manufacturing process have all created awareness of the need to share common information about a product and the process used to make it.

We can go back as far back as the 1950s to find activities which today can be related to the PDES program. The objectives, though changed with viewpoints of technologies, are very similar. The objective of automating the design and manufacturing process is to achieve the highest efficiency and create the highest quality products at the lowest possible prices.

That, again, is our goal with the PDES program. The PDES activity is addressing the following items: the need for exchanging data in a global manufacturing environment; exchanging data between domestic and international trading partners; exchanging data between manufacturing organizations within a corporation, between a corporation and its suppliers, and between a corporation and its customer.

While related technologies and the technology foundation has changed since the 1950s, we have very much the same objectives as we had then. With new technology we are able to address them again to a more complete stage. We find the PDES program to be addressing the identification, categorization, storage, exchange, and sharing of product data throughout the entire product life cycle from its early conceptual design through release to manufacturing, through delivery to the customer, and on through product support.

ARCHITECTURE FOR IMPLEMENTATION

PDES will be implemented to satisfy two primary needs: exchanging product data between different applications, and sharing product data between several different applications.

The PDES organization has defined four implementation architectures; two for exchange, and two for product data sharing.

Level I passive file exchange represents an implementation architecture where product data is transferred from one application to another in a batch process as seen in Figure 1. Level I is similar to the way in which IGES is implemented today. Application A creates product data in a file format as prescribed by the PDES organization. That file is then transferred to Application B. Application B has a post processor translator to post process that file into its own native data structure. This transfer can be bi-directional if both applications have a pre- and post-processor. Level I is focused on the exchange of product data between two applications. Level II has the same focus.

Level II active file exchange is represented by an application that creates product data in a neutral format. This process is outlined in Figure 2. The neutral format has a working form which is a memory resident model of the product data. The two applications, A and B, can reach in to that memory resident model or working form model, and pull out entity by entity the product data it needs. In the exchange process for a Level II implementation, there are two representations of the product data; the exchange file format, or physical format, and another physical form called the working form, which is a memory resident model that can be accessed entity by entity instead of in the entire file.

An example of the Level II implementation is the Air Force PDDI (Product Definition Data Interface) program or GMAP (Geometric Modeling Applications Interface) program. Both programs used a neutral form in a physical file format, and a neutral form in a working form memory resident model format. Both Level I and Level II implementation architectures are focused on the exchange of product data between applications. Level III and IV are focused on the sharing of product data as seen in Figure 3 and 4.

Level III shared database implementation has a logical description of all the product data and a physical implementation of this information in a database format. Applications can then reach into that database and pull out the information they need to drive their specific application. An example would be a relational database with several applications driving from it. Using the data dictionary, which describes data in the database, the applications then have the capability to ask the database manager for pieces of product data it needs to do its job and can provide back end resultant data to be stored into the product definition database. The Air Force integrated design system, or IDS system, developed by Rockwell International, offers a good example of this type of technology.

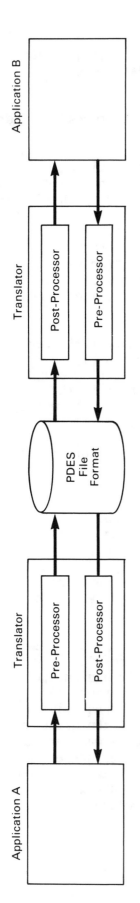

Figure 1. Level I implementation – passive file exchange.

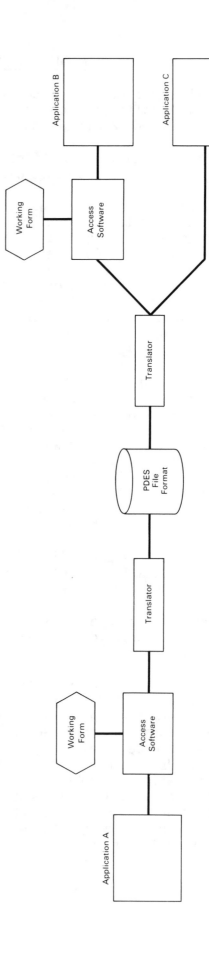

Figure 2. Level II implementation – active file exchange.

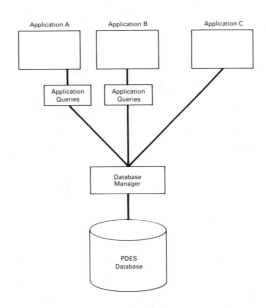

Figure 3. Level III implementation - shared database.

A Level IV knowledge-base implementation of PDES is represented by using an object-oriented approach to the storage of product definition data where the product data and methods to operate on that product data are stored together as an object in an object-oriented database. This knowledge-base representation is the desired implementation environment, and is a limited prototype implementation of a Level IV as it is performed today. Currently, Level IV is not widely implemented throughout industry and the vendor community.

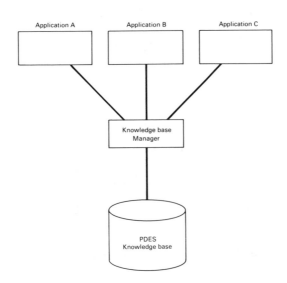

Figure 4. Level IV implementation - knowledgebase.

PDES INC. PROGRAM

The PDES Inc. program is an industry-funded cooperative project to develop the PDES specification and supporting software. The objective of PDES Inc. is to accelerate the development and implementation of PDES within industry. The program is using a combination of industry provided personnel knowledgeable in PDES technologies, and hired contractors working together focusing on the development of the PDES specification.

The background of the PDES program dates back to the fall of 1986. Individuals from a few aerospace companies worked with the United States Air Force Manufacturing Technology Division at Wright Patterson Air Force Base to develop plans for a program to accelerate the development of PDES. Looking at several alternative methods for funding the overall development program, these companies decided to focus on an industry-only membership, where government participation would be in a non-funded involvement.

This activity grew until the development of a interim PDES Inc. planning activity, formulating the initial technical development plan for the PDES Inc. Program, and managing the release of a request for proposal for a PDES Inc. host contractor.

The South Carolina Research Authority teamed with Battelle Memorial Laboratories, Dan Appleton Company, International Technegroup Inc., and Arthur D. Little, to create a proposal to be the host contractor and to provide technical resources for the management of the technical work to be done on the PDES Inc. program. The PDES Inc. board of directors awarded the contract to the South Carolina Research Authority and its technical team in August, 1988. The contract officially was kicked off in October of that year, with a team meeting of the initial professionals from each of the member companies and technical subcontractors involved in the project.

The PDES Inc. program is divided into two major phases, each lasting 18 months. In phase one, the objective is to develop a Level I or Level II implementation of PDES. In phase two, the objective is to develop a Level III implementation of PDES.

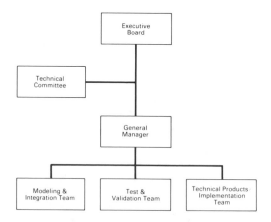

Figure 5. PDES, Inc. organizational structure.

The team of people divided itself into three working groups: the model integration team, the model test and validation team and the technical products/implementation team. This is illustrated in Figure 5. In each group, a team leader was chosen from one of the member companies. A technical subcontractor, and personnel from the different member companies were also assigned to each group.

An important philosophy expressed by the PDES Inc. program defines the relationship of its proprietary results with the PDES volunteer organization. In the spirit of cooperation, PDES Inc. will provide suggested changes to information models received from the volunteer group. The changes will be generated by the PDES Inc. process of modeling, integration, test, validation and implementation of PDES data models. The process and relationship are depicted in Figure 6.

In addition to the three working groups, there was a configuration control board, and a systems integration board comprised of the host contractor general manager, the team leaders from the working groups, and the technical subcontractors. An important link was established between the PDES Inc.

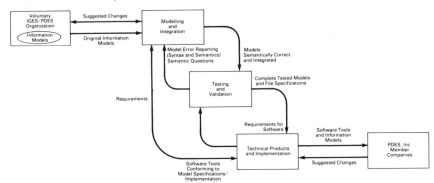

Figure 6. PDES Inc. relationship with PDES voluntary organization.

program and NIST National PDES testbed. NIST became the PDES test laboratory and provided personnel to the PDES Inc. program to help with testing and implementation. NIST received a contract from the Department of Defense to provide these types of services by establishing the PDES National Test Laboratory.

A company can choose to become a member at one of three levels: A Class I member, which is the highest class, pays $100,000 per year, plus provides two technical people for the working groups. In addition, there is a $50,000 travel and equipment budget provided to each of the technical people. Class I members have membership on the board of directors as voting members, and own all rights to all deliverables in the PDES Inc. program.

A Class II member pays $50,000 a year, provides one technical person to the working groups, has a seat on the board of directors in a non-voting capacity, and has the same rights of ownership to the results of the PDES Inc. program.

The third class is an observer class. The fee is $25,000 a year. There is no requirement to contribute technical labor, but there is also no ownership in the results of the PDES Inc. program.

The following table shows the membership as of September, 1989.

Class I	Class II	Class III
Boeing	LTV Aerospace	Honeywell
General Dynamics	Rockwell	
General Electric	Prime Computers	
Grumman	DEC	
Lockheed	FMC	
McDonnell Douglas	Westinghouse	
Northrop	Newport News Shipbuilding	
IBM		
Martin Marietta		
General Motors		
United Technologies		

Table 1. PDES Inc. program membership.

The PDES Inc. program will deliver results in a time frame which can be incorporated into the CALS Phase II effort. The Department of Defense, although not funding the PDES Inc. program outside of funding NIST, is supporting the PDES Inc. program by providing projects to companies that are involved in PDES Inc. The projects require the use of PDES in the delivery of data to the government. They are also sending encouragement to companies who have not yet joined the PDES Inc. program.

The planning people in charge of the CALS activity at the Department of Defense see the PDES Inc. program as providing PDES results to be used as the cornerstone of the data content and data delivery in a CALS Phase II environment for the future.

THE BENEFITS OF PDES

Looking at manufacturing from a bottom-up functional view, we find the introduction of computer technology can automate several of those discrete functions. Each implementation of automation itself becomes a small island of automation. An automated function typically has involved with it a user interface, some kind of functionality, or algorithm coded into a software program, data input, data output, and data storage. The data repository that becomes the culmination of running this automated function several times becomes a small database.

The spectrum of manufacturing functionality from design through manufacturing to product support, has a great deal of information about our business stored in discrete implementations of technology. Implementations that are not interchangeable and cannot be integrated, can reside on top of other databases, or use information generated by other automated functions.

This is the problem we face today as we attempt to implement new technologies that may change a particular functional application. Every time we do this, we face investment in changing our technology. This is expensive and time consuming.

As an industry, we need to store information in a form we can retrieve year after year and still use, even if it is in a different form of automated functions. With the rate of change we have today in technology, we are

160

unable to accomplish this. We cannot, for example, come back in five years with the same CAD/CAM system and expect to retrieve the same engineering model that we had. The technology changes too fast. With new technology, we eliminate paper. We have paper engineering drawings that are 100 years old. Paper has been an outstanding form of interface between various functional applications. This is because a human was involved with each functional applications.

Now, we have automated applications, computer-aided design/computer-aided manufacturing, engineering analysis, all of these are done on computer. Inspection, too, is done with computer-driven machine tools. Automated assembly and automated material handling have as a function some computer automation, which requires information input and generally creates information output.

The total sum of this information is what we have defined earlier as product data; data that defines the product and the process used to manufacture it.

We've made attempts before to define neutral formats for storage and retrieval. One significant attempt is IGES. And, although IGES is used very heavily in the exchange of engineering information between two different CAD systems, it has not been successful in long-term storage and retrieval. IGES simply was not designed to do that. It contains a file structure and a format which does not provide easy storage. The PDES organization needs to be considering that. PDES needs to be addressing the implementation levels that we have discussed between exchange and sharing of information. It also needs to be addressing the long-term storage and retrieval of product data.

The importance of PDES from a business viewpoint is that it will provide not only a format for storing, archiving, and exchanging information, but the definition of that information. This will allow us to create automated applications which link directly to the PDES structure, and which can use PDES information to perform their functional capability.

It is important for those of us involved in the development and use of PDES to recognize the limitation and scope of PDES. PDES as a technology will provide the capability to exchange and share information. PDES as a technology does not answer all the questions faced by information management. While PDES addresses the information flow into and out of all the functional applications within a company, it does not address the organization of the company's handling of information, nor does it address the management of information within the company.

It is up to the manufacturing professional to address the capability of handling information within individual companies. The format and content of that information can now be changed because of the introduction of PDES. With PDES, we can handle the information in an electronic form, define enough structure into it to fill it with as much intelligent product and process information as possible, and still drive all the different functional applications in the manufacturing enterprise.

The information management within your own business may yet become another functional application within your manufacturing enterprise. As you approach what has been called a computer integrated enterprise, capabilities provided by a PDES specification and PDES implementation software will assist you in implementing new technology to achieve an overall integrated enterprise.

Part Six

CIM Survey

Survey of World-Class CIM Planning: Information Architecture First

by Frederick J. Michel
BMD International

Mark D. Pardue
Old Dominion University

INTRODUCTION

In analyzing data from a wide range of companies that had successfully implemented CIM, we focused on eight world-class Fortune 500 companies. We identified several factors important to their success with CIM. These companies all shared the following attributes:

* A multidisciplinary team for CIM strategic planning composed of management, engineering, manufacturing and shop floor personnel.
* Top-level CIM architecture, with a set of hardware, software and communications standards applied across the company but allowing flexibility for individual plant requirements.
* The commitment and understanding of senior management.
* A strategic, long-term view.
* Adaptation of a contribution-based accounting system.
* Implementation of a CIM training program for all levels of the company.
* Partnership with suppliers.
* Establishment of a corporate CIM organization.

From a literature search of 27 trade publications, we identified 76 companies that had performed successful CIM strategic planning or had implemented 10 or more CIM related technologies as shown in Figure 1. We then identified several leading candidates for study in the area of CIM strategic planning and obtained interviews with key senior management officials at eight of them. All were Fortune 500 corporations. These included manufacturers of electrical components for military and commercial buyers, a machine tool manufacturer, a maker of agricultural equipment, and manufacturers of weapons systems and aircraft.

The survey found that the primary motivation for the implementation of CIM was survival of product lines in the face of increasing cost and quality competition, and in some cases, survival of the company itself. CIM was viewed as a way to cut costs, improve product quality, shorten delivery times, and provide flexibility. For most companies, however, the risks of implementing CIM seemed to overshadow these potential gains unless survival was at stake, at which point the risks became moot; failure with CIM would simply accompany the failure of an already moribund product line. The fallacy here is that if companies wait until they are in decline to

CIM Area or Technology	Number (and Percentage) of Companies*	
CAD/CAE/CAM	39	(37%)
Enterprise-Integrated Information Systems	31	(29%)
Just-in-Time Manufacturing	29	(27%)
Manufacturing Automation	28	(26%)
Robotics	28	(26%)
Computer or Direct Numerical Control Machines	25	(24%)
Automated Materials-Handling Systems	25	(24%)
Flexible Manufacturing Systems	23	(22%)
Shop Floor Scheduling Systems	21	(20%)
Factory Communications Systems	19	(18%)
Materials Requirements Planning	18	(17%)
Coordinate Measuring Machine or Optical Inspection	16	(15%)
Bar Coding and Automated Parts Identification	12	(11%)
Total Quality Management	12	(11%)
Artificial Intelligence and Expert Systems	8	(8%)
Statistical Process Control	8	(8%)
Computer-Automated Process Planning	8	(8%)
Concurrent Engineering	6	(6%)
Group Technology	5	(5%)
Manufacturing Simulation	5	(5%)

Note:
*The total sample consisted of 106 companies; however,
many companies implemented more than one area or technology.

Figure 1. Number of companies identified through literature search implementing key CIM areas and technologies.

implement CIM, their resources are low. CIM should instead be implemented as an integral part of strategic planning.

PLANNING TEAM PARTICIPANTS AND FINANCIAL ANALYSES

Figure 2 lists the strategic planning steps used by the selected companies. Each respondent indicated that its CIM strategic planning team was made up of representatives from all departments, including representatives from the corporate CIM organization. For the most part, no outside expertise was used. The planning team reported to the CEO, the vice president for manufacturing, a CIM steering committee, or in the case of two defense contractors, the director of the CIM organization.

All respondents performed some type of financial analysis for their CIM projects, but most did not use the results from these analyses as critical factors to justify a CIM project. Two companies indicated that payback was important except for enabling projects; that is, projects that provide a basis on which to implement other CIM projects. Defense related companies usually performed cost/benefit analyses; other companies calculated return on investment, payback, or the total economic impact on the company. Except for companies receiving Department of Defense Industrial Modernization Incentives Program (IMIP) funds to subsidize their CIM efforts, all respondents stated that CIM projects are long-term investments and that calculations of saving must include the improved quality, throughput, and flexibility attained with CIM as well as greater customer satisfaction and an enhanced ability to meet strategic objectives.

All respondents indicated that representation by all departments and the commitment of senior management heightened cooperation, participation, and enthusiasm from everyone involved. Another leading response was that CIM justification was based on strategic objectives and mission needs, with cost justification secondary, at least for enabling CIM projects.

Companies generally begin CIM implementation with a pilot program; the respondent's consensus was that the area piloted first should be strategically important and should have the potential for early success.

Company	Steps in Planning Process	Type of Process
A	Identify: —Mission and Scope —Objectives —Risks —Strategic issues —Benefits of CIM —Specific implementation strategies —Management by objectives	Top-down
B	Information withheld	Top-down/bottom-up
C	List projects Rank projects* Determine cost of each project Narrow down the list of projects	Top-down/bottom-up
D	Identify: —Mission —Strategy —Future business —Cost drivers* —Architectures* —Integrate projects*	Top-down/bottom-up
E	Information withheld	Top-down/bottom-up
F	Understand business enterprise* Eliminate unnecessary functions* Consolidate functions* Create a detailed CIM plan*	Top-down/bottom-up
G	Analyze operations* Perform computer modeling Use group technology to integrate design* Integrate information systems architecture* Integrate information control systems Use CIM factory design	Top-down/bottom-up
H	Organize a multidisciplinary project planning team* Develop the manufacturing strategy* Use systems design approach* Evolve cost-effective implementation* Merge manufacturing and engineering* Use closed-loop control*	Top-down/bottom-up

Note:
*Steps company considers critical

Figure 2. Steps in the CIM strategic planning process.

KEY IMPLEMENTATION AREAS

CAD/CAE/CAM has been by far the most prevalent CIM technology implemented. Two companies that implemented this as their first CIM project, however, noted that the benefits were marginal and that these benefits result from improved drafting productivity. Only after integration across the entire enterprise took place were the real benefits realized. These two companies stated that in retrospect, they would not select CAD/CAE/CAM as the pilot program. Initial projects recommended by the respondents were:

* Interdepartmental communications.
* An enterprise-integrated database.
* An enterprise wide information system.
* Group technology.
* Computer-aided manufacturing.

Three of these five recommended areas involve coordinated information sharing across the entire enterprise. The implication is that establishing a top level architecture is considered an important first step in a company's CIM implementation.

Three of the companies most successful in implementating CIM merged the product-engineering and manufacturing-engineering groups. This reorganization supports simultaneous engineering and avoids the "throwing-it-over-the-wall" syndrome where the groups have little communication with each other, and engineers simply forward finished designs to manufacturing. A company that had not merged product engineering with manufacturing engineering noted coordination problems between the two groups. All respondents established a separate corporate CIM group. At one company, this group implements CIM directly, whereas at other companies, the user divisions have that responsibility. At two companies, the vice president of manufacturing and operations has this responsibility.

Another change mentioned was not organizational, but involved using shop floor personnel to help engineers plan and implement CIM.

CIM training programs are usually implemented throughout the organization, from shop floor to CEO, but most companies do not bring in outside expertise to plan and implement CIM, except for such specialized areas as statistical process control, in which external contractors are used. A lead engineer is usually responsible for becoming expert in the CIM technology the company wishes to implement. At three companies, CIM implementation resulted in product design changes to improve manufacturability. The other company stated that CIM did not result in design changes.

The relationships that companies have with suppliers change with CIM implementation; they become closer and involve fewer suppliers. Requirements for higher quality and shorter delivery times necessary to support Just-in-Time (JIT) manufacturing are passed on to the suppliers. Through CIM, companies come to see suppliers as venture partners.

All respondents on the subject of accounting indicated that the standard labor-burden accounting methods are unsatisfactory for CIM implementation and that costing methods based on the contribution of each activity must be used. Some companies have already changed their accounting systems along these lines; others stated that such changes are forthcoming.

All respondents indicated that they had some type of quality program, and the CIM was a significant factor in improving product quality. CIM reduces rework, scrapping, and the number of inspectors. In the opinion of most respondents, CIM ensures that quality is built into the product and eliminates the need for a quality assurance department. Some companies place the responsibility for inspection of purchased parts with the parts suppliers, whereas other companies use the CIM database to automate the inspection of incoming parts. When placing inspection responsibility with the supplier, the company usually continues sample inspections of incoming purchased parts until the quality performance of the supplier can be certified, sometimes on a parts-number basis. Many companies indicated that in-process inspection of parts has been automated using specification data in the CIM database, and is an integral part of the manufacturing process.

Most companies indicated that they standardized CIM hardware, software and communications across the company. The standard they chose either already had the flexibility to let the company address individual plant problems, or was modified as required. Among the respondents, however, there was no discernible preference for a particular hardware or software. A variety of communications network standards were used, including the Manufacturing Automation Protocol (MAP).

The most common implementation barrier mentioned by the respondents was the lack of capital for all of the needed CIM projects. Other problems and corresponding solutions are shown in Figure 3.

Problem	Solution
Lack of technical talent to plan and implement CIM	Train your own employees
Lack of coordination between departments	Form CIM teams with representatives of all departments
	Merge product engineering and manufacturing engineering
Technical problems on CIM projects	Keep management informed
	Postpone production start-up date until technical problems have been solved
Resistance to change	Give educational presentations for management
	Involve everyone as part of the CIM strategic planning team
Absence of software for planning the factory based on group technology concepts	Develop software in-house
Lack of attention and commitment by total organization	Encourage support and commitment from every corporate level
	CIM organization must continuously address this problem
Lack of a CIM master plan	Have corporate CIM organization develop a master plan using project charters and milestone reports

Figure 3. CIM implementation problems and solutions.

Several respondents gave an indication of the resulting cost/benefit or payback of CIM. Payback periods were from two to seven years. The companies also indicated difficulties in quantifying benefits. Three-forths of the companies met or exceeded their original projections. All eight selected companies either remained competitive or improved their competitive position by reducing costs, improving quality and reducing delivery times to customers.

CASE HISTORIES

We interviewed three leading companies in the CIM strategic-planning area. Each company's experience illustrates a significant aspect of CIM

COMPANY A

For Company A, survival of one product line was the initial motivation for considering CIM. The company determined that the only solution to overwhelming competitive pressures in this product line was to cut labor costs, and that CIM would be the most effective way to do this.

Company A's strategic CIM-planning process is applied only to new products and consists of a formal, documented process identifying the following:

* The mission and scope.
* The objectives.
* Benefits of CIM.
* Risks.
* Strategic issues.
* Specific implementation strategies.
* Management by objectives.

The development of Company A's CIM strategic plan works from the top down, with the participation of the executive group (especially the CEO), the quality and operations departments, and each product group; no outside expertise is used in the planning process. Managers and directors develop the plan through an iterative process. The leader of the planning team reports to the executive vice president of the company. Supporting financial analyses start with marketing estimates; then target product-cost analyses are performed, and return on investment (ROI) is calculated using the entire product life as the time base. The priority given to CIM projects is based on strategic importance, not financial attractiveness.

To support CIM implementation, Company A combined its engineering services and manufacturing organizations. This new manufacturing operations group implements CIM projects and one of its main goals is to design new products for maximum producibility. The ultimate responsibility rests with the vice president of operations.

CIM allows the company to automate testing. Inspection of purchased parts is the responsibility of the suppliers through a certification program (certification is based on parts numbers). Inspection of in-house manufactured parts is the responsibility of the machine operator in accordance with the company's quality assurance program. As specifications become tighter on the finished product, Company A works with its parts suppliers to get agreement on critical dimension specifications, which have been formalized through the CAD/CAE/CAM database. JIT manufacturing requirements have also been passed on to the suppliers.

Little training in CIM has occurred within the company. Statistical process control training has been performed across all management levels by a combination of in-house personnel and outside contractors.

The accounting system was modified to support CIM implementation. For the new products manufactured by the CIM system, the traditional allocation of indirect cost as a burden on direct labor was changed to a contribution method. Implementation priority is based on strategic considerations and mission needs; cost is not a primary factor.

The particular problems encountered by Company A in CIM implementations were the lack of technical talent, finances, and coordination.

The first two problems still exist. Coordination, however, was improved through group meetings and the company is developing an architecture to assist in coordination.

The first CIM project for Company A cost approximately $15 million. The company standardized on Sun hardware and Daisy software. Since the new facility was for a new product line, there was no basis for determining the cost saving. However, the company stated that the payback period from the very first expenditure is seven years. The company generally does not perform detailed financial analyses, so no comparison with original projections is possible. Based on their initial success with the facility, however, the company is continuing CIM implementation with new product lines.

The biggest effect on the company's way of doing business is that hourly workers on the factory floor can now suggest possible improvements to the engineers.

COMPANY B

After preliminary CIM efforts on a very small scale with correspondingly insignificant results were implemented after a decrease in sales, Company B was forced to start a full-scale CIM project. CIM was viewed as a way for the company to reduce its costs and its break-even point. The representative from Company B felt that companies will not undergo the pain of change required to implement CIM unless their survival is threatened, and indicated that there is no real overall CIM strategic planning process in use at Company B. Each factory selects and plans its own CIM project, and the corporate CIM staff provides support as requested. The corporate CIM staff may also propose projects to a factory, but it is the prerogative of the factory to initiate them. Each project planning team is multidisciplinary and includes technology experts, functional area experts and management. In some cases, outside expertise is used, based on the factory's needs. Each project is justified on the basis of a return on investment analysis.

Company B has established a centralized corporate staff of CIM experts that the factories call as needed on a project-by-project basis. However, the factory manager has the ultimate responsibility for CIM project implementation.

Company B management feels that two key factors in successful CIM implementation are the use of group technology organized around product families, and the reduction in the size of departments. The company has also made organizational changes by assigning the engineering director atsome units the responsibility for both product engineering and manufacturing engineering. The company uses recommendations of shop floor personnel for redesigning tools, rearranging fixtures and changing manufacturing methods. Product design changes are made as needed to accommodate CIM. The burden of inspecting purchased parts is placed on the suppliers and inspection of manufactured parts is the responsibility of the manufacturing operations. Statistical process control is used to some extent so fewer inspectors are needed. The company also uses fewer suppliers, but quality and delivery requirements have been tightened on those suppliers.

Company B has developed a new activity based accounting system for CIM that, rather than using the standard labor-burden cost accounting approach, determines what activities are required for a product from concept through sale, and associates costs with each activity to determine product cost. Group technology is considered the key CIM technology to implement for simplifying before automating, which is a company principle. Automated materials handling and JIT techniques can then be implemented from the more focused work cell approach made possible with the use of group technology.

The main problem in Company B's CIM implementations has been a lack of qualified technical people. The company was forced to train its own people in the use and application of CIM technology. Nevertheless, in most cases, the results compared favorably with the original projections. One group technology project alone resulted in the reduction of parts from 405 to 70, an 80% reduction in scrap, a 42% reduction in cycle time, and an inventory reduction from 21 to 10 days, all within a two year period. The company has followed up its initial success by streamlining, using group technology on a factory-by-factory basis. Throughout the company, one vendor's product offerings have been standardized, with heterogeneous computer equipment tied together with MAP. This lets the company tailor solutions to a particular plant while retaining a company wide concept standard.

COMPANY C

Survival of the company and an effort to excel were the primary motivations for Company C to implement CIM. Through the leadership of its CEO, the company looked toward improving its competitive position during the next 10 years.

The planning process at Company C is driven by the CEO but functions as a bottom-up, cooperative process between the users in the various departments and the systems group. Outside expertise was brought in when specialized knowledge not existing in the company was needed. The systems group and the users jointly develop a wish list of projects, with the departments ranking their projects according to their needs. The systems group develops the cost and savings estimates for each project; the savings estimates are developed through the use of a comprehensive economic model of the company's entire business enterprise.

The project list is reviewed in a group in which department vice presidents defend their projects; the group narrows the project list to bring it within the range of available resources. Although this process can get heated, few conflicts require intervention of the CEO for resolution.

Throughout the process, projects are ranked according to user needs; cost reductions are not a critical factor in the selection of projects. Company C believes that basing user priorities on need is a key step in planning. Users have the ultimate responsibility for CIM implementation, assisted by the systems group, and the user who proposes a CIM project becomes the group manager for that project.

Company C is also merging engineering and manufacturing. This move is facilitated by the use of a company wide database shared by both engineering and manufacturing. The company's first CIM project was CAD/CAM, because at the time it had not established a way of assigning priorities. The company believes that a central database is an important first step in implementing CIM because it allows control of the entire business operation. The same database is also used for recording inspection data on critical dimensions to enhance the effectiveness of coordinate measurement machines. The database has enabled the company to reduce tolerances on its systems, which also resulted in reduced tolerance requirements for parts suppliers. Furthermore, the company has reduced leadtimes from its suppliers.

The systems group personnel who design each new CIM module develop a special training program for the users. These training programs are administered by the systems group, and are given to all company personnel from CEO to shop floor workers.

The company's accounting system has been modified to support CIM implementation. Machine hours, instead of worker hours, are used in calculating burden for operations that require little labor. Company management believes that the accounting system will have to be modified further.

As problems arose in CIM projects, the systems group developed solutions at the expense of start up dates to ensure trouble-free systems operation upon implementation.

For the CAD/CAM project--the company's first CIM implementation--the systems group used its comprehensive economic model to project a cost saving of $1 million per year beginning with the sixth year. The actual saving after five years was $1.014 million. Since this initial success, the company has been working on a completely integrated information system.

On the basis of this experience, the company expects future projects to realize paybacks within seven or eight years. One soft benefit of CIM implementation that Company C has realized is increased quality.

Index

TOP, See: Technical and Office
   Protocol
Training, 30, 56, 70
Transport control, 59
Transport profiles, 60
Turnkey vendors, 72
Two-way communication, 30

U

Universal work process
   language, 47
User groups, 69, 72

V

Vendor training, 70
Vendors, 51, 54, 61, 63, 64

W

Work cells, 45
Work stations, 26
Work-in-process, 18, 26
Worker skill level, 20